D1539232

The Handbook
of Environmental Chemistry

Volume 4   Air Pollution
Part F

O. Hutzinger
Editor-in-Chief

Indoor Air Pollution

Volume Editor: Peter Pluschke

With contributions by
G. A. Ayoko · K. Balakrishnan · P. Bluyssen
W. Butte · J. M. Hao · R. Keller · L. Morawska
P. Ramaswamy · H. Rüden · T. Salthammer
S. Sankar · H. Schleibinger · B. A. Tichenor
T. L. Zhu

Springer

Volume Editor

Dr. Peter Pluschke
Chemisches Untersuchungsamt
Postfach
90317 Nürnberg
Germany
peterpluschke@odn.de

Programme de la Gestion et de la Protection
de l'Environment (PGPE)
c/o Bureau de la GTZ
B. P. 433
1000 Rabat R. P.
Maroc

ISSN 1433-6855
ISBN 3-540-21098-9
DOI 10.1007/b82868
Springer-Verlag Berlin Heidelberg New York

Library of Congress Control Number: 2004104243

Bibliographic information published by Die Deutsche Bibliothek
Die Deutsche Bibliothek lists this publication in the Deutsche Nationalbibliographie;
detailed bibliographic data is available in the Internet at <http://dnb.ddb.de>.

Springer-Verlag is a part of Springer Science+Business Media GmbH
springeronline.com
© Springer-Verlag Berlin Heidelberg 2004
Printed in Germany

Production Editor: Christiane Messerschmidt, Rheinau
Cover Design: E. Kirchner, Springer-Verlag
Typesetting: Fotosatz-Service Köhler GmbH, Würzburg
Printed on acid-free paper   52/3020 – 5 4 3 2 1 0

The Handbook of Environmental Chemistry
Also Available Electronically

For all customers who have a standing order to The Handbook of Environmental Chemistry, we offer the electronic version via SpringerLink free of charge. Please contact your librarian who can receive a password for free access to the full articles by registering at:

springerlink.com

If you do not have a subscription, you can still view the tables of contents of the volumes and the abstract of each article by going to the SpringerLink Homepage, clicking on "Browse by Online Libraries", then "Chemical Sciences", and finally choose The Handbook of Environmental Chemistry.

You will find information about the

– Editorial Bord
– Aims and Scope
– Instructions for Authors
– Sample Contribution

at springeronline.com using the search function.

Preface

Environmental Chemistry is a relatively young science. Interest in this subject, however, is growing very rapidly and, although no agreement has been reached as yet about the exact content and limits of this interdisciplinary discipline, there appears to be increasing interest in seeing environmental topics which are based on chemistry embodied in this subject. One of the first objectives of Environmental Chemistry must be the study of the environment and of natural chemical processes which occur in the environment. A major purpose of this series on Environmental Chemistry, therefore, is to present a reasonably uniform view of various aspects of the chemistry of the environment and chemical reactions occurring in the environment.

The industrial activities of man have given a new dimension to Environmental Chemistry. We have now synthesized and described over five million chemical compounds and chemical industry produces about hundred and fifty million tons of synthetic chemicals annually. We ship billions of tons of oil per year and through mining operations and other geophysical modifications, large quantities of inorganic and organic materials are released from their natural deposits. Cities and metropolitan areas of up to 15 million inhabitants produce large quantities of waste in relatively small and confined areas. Much of the chemical products and waste products of modern society are released into the environment either during production, storage, transport, use or ultimate disposal. These released materials participate in natural cycles and reactions and frequently lead to interference and disturbance of natural systems.

Environmental Chemistry is concerned with reactions in the environment. It is about distribution and equilibria between environmental compartments. It is about reactions, pathways, thermodynamics and kinetics. An important purpose of this Handbook, is to aid understanding of the basic distribution and chemical reaction processes which occur in the environment.

Laws regulating toxic substances in various countries are designed to assess and control risk of chemicals to man and his environment. Science can contribute in two areas to this assessment; firstly in the area of toxicology and secondly in the area of chemical exposure. The available concentration ("environmental exposure concentration") depends on the fate of chemical compounds in the environment and thus their distribution and reaction behaviour in the environment. One very important contribution of Environmental Chemistry to the above mentioned toxic substances laws is to develop laboratory test methods, or mathematical correlations and models that predict the environmental fate of new chemical compounds. The third purpose of this Handbook is to help in the basic understanding and development of such test methods and models.

The last explicit purpose of the Handbook is to present, in concise form, the most important properties relating to environmental chemistry and hazard assessment for the most important series of chemical compounds.

At the moment three volumes of the Handbook are planned. Volume 1 deals with the natural environment and the biogeochemical cycles therein, including some background information such as energetics and ecology. Volume 2 is concerned with reactions and processes in the environment and deals with physical factors such as transport and adsorption, and chemical, photochemical and biochemical reactions in the environment, as well as some aspects of pharmacokinetics and metabolism within organisms. Volume 3 deals with anthropogenic compounds, their chemical backgrounds, production methods and information about their use, their environmental behaviour, analytical methodology and some important aspects of their toxic effects. The material for volume 1, 2 and 3 was each more than could easily be fitted into a single volume, and for this reason, as well as for the purpose of rapid publication of available manuscripts, all three volumes were divided in the parts A and B. Part A of all three volumes is now being published and the second part of each of these volumes should appear about six months thereafter. Publisher and editor hope to keep materials of the volumes one to three up to date and to extend coverage in the subject areas by publishing further parts in the future. Plans also exist for volumes dealing with different subject matter such as analysis, chemical technology and toxicology, and readers are encouraged to offer suggestions and advice as to future editions of "The Handbook of Environmental Chemistry".

Most chapters in the Handbook are written to a fairly advanced level and should be of interest to the graduate student and practising scientist. I also hope that the subject matter treated will be of interest to people outside chemistry and to scientists in industry as well as government and regulatory bodies. It would be very satisfying for me to see the books used as a basis for developing graduate courses in Environmental Chemistry.

Due to the breadth of the subject matter, it was not easy to edit this Handbook. Specialists had to be found in quite different areas of science who were willing to contribute a chapter within the prescribed schedule. It is with great satisfaction that I thank all 52 authors from 8 countries for their understanding and for devoting their time to this effort. Special thanks are due to Dr. F. Boschke of Springer for his advice and discussions throughout all stages of preparation of the Handbook. Mrs. A. Heinrich of Springer has significantly contributed to the technical development of the book through her conscientious and efficient work. Finally I like to thank my family, students and colleagues for being so patient with me during several critical phases of preparation for the Handbook, and to some colleagues and the secretaries for technical help.

I consider it a privilege to see my chosen subject grow. My interest in Environmental Chemistry dates back to my early college days in Vienna. I received significant impulses during my postdoctoral period at the University of California and my interest slowly developed during my time with the National Research Council of Canada, before I could devote my full time of Environmental Chemistry, here in Amsterdam. I hope this Handbook may help deepen the interest of other scientists in this subject.

Amsterdam, May 1980   *O. Hutzinger*

Twentyone years have now passed since the appearance of the first volumes of the Handbook. Although the basic concept has remained the same changes and adjustments were necessary.

Some years ago publishers and editors agreed to expand the Handbook by two new open-end volume series: Air Pollution and Water Pollution. These broad topics could not be fitted easily into the headings of the first three volumes. All five volume series are integrated through the choice of topics and by a system of cross referencing.

The outline of the Handbook is thus as follows:

1. The Natural Environment and the Biochemical Cycles,
2. Reaction and Processes,
3. Anthropogenic Compounds,
4. Air Pollution,
5. Water Pollution.

Rapid developments in Environmental Chemistry and the increasing breadth of the subject matter covered made it necessary to establish volume-editors. Each subject is now supervised by specialists in their respective fields.

A recent development is the accessibility of all new volumes of the Handbook from 1990 onwards, available via the Springer Homepage http://www.springer.de or http://Link.springer.de/series/hec/ or http://Link.springerny.com/series/ hec/.

During the last 5 to 10 years there was a growing tendency to include subject matters of societal relevance into a broad view of Environmental Chemistry. Topics include LCA (Life Cycle Analysis), Environmental Management, Sustainable Development and others. Whilst these topics are of great importance for the development and acceptance of Environmental Chemistry Publishers and Editors have decided to keep the Handbook essentially a source of information on "hard sciences".

With books in press and in preparation we have now well over 40 volumes available. Authors, volume-editors and editor-in-chief are rewarded by the broad acceptance of the "Handbook" in the scientific community.

Bayreuth, July 2001   *Otto Hutzinger*

Contents

The Handbook of Environmental Chemistry Vol. 4, Part F (2004): 1–35
DOI 10.1007/b94829
© Springer-Verlag Berlin Heidelberg 2004

Volatile Organic Compounds in Indoor Environments

Godwin A. Ayoko (✉)

International Laboratory for Air Quality and Health, School of Physical and Chemical
Sciences, Queensland University of Technology, GPO 2434, QLD 4001, Australia
g.ayoko@qut.edu.au

Abstract   This chapter provides an overview of the types, sources and current techniques for characterising volatile organic compounds (VOCs) in nonindustrial indoor environments. It reviews current knowledge on the levels of VOCs in indoor environments, discusses concepts for regulating indoor levels of VOCs and appraises current efforts to understand the links between VOCs and building-related health/sensory effects. It also provides an up-to-date outline of new trends in and perspectives for indoor air VOC research.

Abbreviations

AFoDAS/AVODAS	Automated formaldehyde data acquisition system/automated volatile organic compounds data acquisition system
ECA	European Collaborative Action
ECD	Electron capture detector
ETS	Environmental tobacco smoke
EXPOLIS	Air pollution exposure distributions of adult urban populations in Europe
FID	Flame ionisation detector
GC	Gas chromatography
HPLC	High-performance liquid chromatography
IAQ	Indoor air quality
MS	Mass spectrometry
PAS	Photoacoustic spectroscopy
PDMS	Poly(dimethylsiloxane)
SBS	Sick building syndrome
SER	Area-specific emission rate
SPME	Solid-phase microextraction
SSV	Safe sampling volume
SVOC	Semivolatile organic compounds
TOF	Time of flight
TVOC	Total volatile organic compounds
US EPA	United States Environmental Protection Agency
VOC	Volatile organic compounds
VVOC	Very volatile organic compounds

1
Introduction

There is a long history of interest in volatile organic compounds (VOCs) in indoor environments. This is evidenced by the large number of national and regional studies/campaigns that have been undertaken to model, identify or quantify indoor VOCs or that relate indoor levels of VOCs to indoor materials, indoor activities and some perceived health/sensory effects. The main interest in such studies lies in the fact that most people spend up to 80% of the day in one indoor environment or another, where pollution levels can be higher, pollutant sources are more varied and exposures are more important than those found in outdoor microenvironments. Many novel insights have emerged from the studies, and some of the main features of these insights are outlined in this chapter. In particular, the types of VOCs commonly found in indoor air, sources/source characteristics of indoor VOCs, measurement techniques for profiling indoor VOCs, typical results from indoor air VOC studies, health effects of VOCs, concepts for reducing indoor VOCs and new trends in indoor VOC studies, particularly in the last decade, are discussed in the following sections.

To put the concepts discussed in the chapter in the right context, distinction must first be made among the terms very volatile organic compounds (VVOC), VOCs, semivolatile organic compounds (SVOCs) and particulate organic matter (POM), which are commonly used to describe organic compounds in indoor air. According to the WHO [1], VVOCs, VOCs, SVOCs and POM are compounds with boiling ranges between 0 °C and 50–100 °C, 50–100 °C to 240–260 °C, 240–260 °C to 360–400 °C and higher than 380 °C, respectively.

2
Types of Indoor VOCs

Hundreds of VOCs are found in a typical nonindustrial indoor environment. Many of these compounds are aromatic hydrocarbons, alkenes, alcohols, aliphatic hydrocarbons, aldehydes, ketones, esters, glycols, glycolethers, halocarbons, cycloalkanes and terpenes [2] but amines like nicotine, pyridine, 2-picoline, 3-ethenylpyridine and myosmine are also widespread, especially in smoking microenvironments [3]. Moreover, low molecular weight carboxylic acids, siloxanes, alkenes, cycloalkenes and Freon 11 are frequently encountered in typical nonindustrial indoor air [1].

3
Sources of Indoor VOCs

VOCs are ubiquitous in indoor environments. They are widespread in household and consumer products, furnishing and building materials, office equipment, air fresheners, paints, paint strippers, household solvents and in microorganisms found in indoor environments. In addition, humans and their indoor activities such as cooking, cleaning, building renovation and tobacco smoking generate high levels and wide varieties of VOCs. Apart from these indoor sources, intrusions of VOCs from outdoor traffic as well as biogenic and industrial emissions contribute significantly to indoor VOC levels. Furthermore, indoor air reactions are now recognised as sources of indoor VOCs, as exemplified by the reaction of ozone with 4-phenylcyclohexene in carpets and with latex paints to generate appreciable amounts of aldehydes [4].

While some common indoor VOCs originate exclusively from indoor sources, others have multiple indoor and outdoor sources. Consequently, the indoor level of a particular VOC is the summation of the contributions of its different indoor and outdoor sources. Various authors have undertaken comprehensive reviews of indoor VOC sources [5–9] and it is apparent from these reviews that the main sources of the typical indoor VOCs together with the major VOC chemical classes associated with the sources are as summarised in the following.

- Outdoor sources: Traffic, industry (aliphatic and aromatic hydrocarbons; aldehydes; ketones; esters).
- Building material: Insulation, paint, plywood, adhesives (aliphatic and aromatic hydrocarbons; alcohols; ketones; esters).
- Furnishing material: Furniture, floor/wall coverings (aliphatic and aromatic hydrocarbons; alcohols; halocarbons; aldehydes; ketones; ethers; esters).
- Garage and combustion appliances: Vehicle emission, tobacco smoking, candles (aliphatic and aromatic hydrocarbons; aldehydes, amines).
- Consumer products: Cleaning, personal care products (aliphatic and aromatic hydrocarbons; alcohols; halocarbons; aldehydes; ketones; terpenes; ethers; esters).
- Equipment: Laser printers, photocopiers, computers, other office equipment (aromatic hydrocarbons; aldehydes; ketones; esters).
- Indoor activities: Cooking, tobacco smoking, use of water and solvents (amines; aliphatic and aromatic hydrocarbons; aldehydes; halocarbons).
- Ventilation systems: Filters of heating, ventilation and air-conditioning systems (aliphatic and aromatic hydrocarbons; alcohols; halocarbons; aldehydes; ketones; terpenes; ethers; esters).
- Biological sources: Humans, moulds, bacteria, plants (terpenes, glycoesters; alcohols; esters; aldehydes).

4
Sampling and Characterisation of Indoor VOCs

Interest in indoor air monitoring is driven by a wide variety of reasons [10–11]; the most prominent ones include the desire to

- Undertake baseline measurements in order to set limits.
- Identify the presence of specific pollutants (e.g. formaldehyde).
- Apportion indoor VOC sources.
- Evaluate levels of compliance with legislations.
- Assess contaminated buildings.
- Apply and validate sampling/analysis methods.
- Validate model s.
- Evaluate ventilation systems.
- Evaluate the strength of a specific source.
- Relate sick building syndrome (SBS)/health effects to VOC levels.
- Understand the mechanisms of VOC transport from source to receptor sites.

While specific details may differ, the general analytical procedures described in the following and summarised in Scheme 1 apply to most monitoring exercises. Firstly, the purpose of the monitoring exercise must be clearly set out, then an appropriate method of sampling must be chosen, followed (where applicable) by the choice of suitable methods for sample storage, sample preparation or preconcentration and sample separation. Lastly, identification and/or quantification of the components are performed [12].

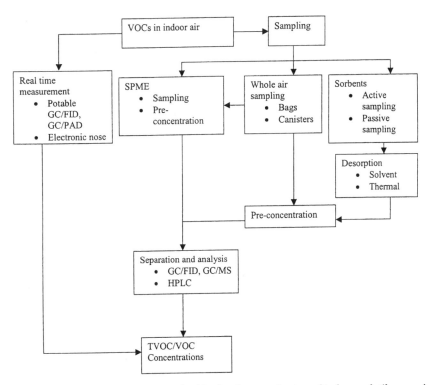

Scheme 1 Summary of the steps involved in the characterisation of indoor volatile organic compounds (*VOCs*)

Assessment of VOCs levels in an indoor microenvironment may be accomplished by direct measurements or by collection of a sample of air followed by subsequent laboratory analysis of the sample. Both of the approaches can be devised to answer the basic questions what is present and how much is present?

4.1
Direct Measurements

In general, the direct measurement is achieved through the use of portable gas chromatography (GC), photoacoustic spectroscopy (PAS), IR spectroscopy and more recently by the so-called electronic noses [7, 12, 13]. Ekberg [14] recently undertook a review of direct reading instruments used in monitoring organic compounds. Such real-time measurement instruments facilitate rapid data acquisition and are especially useful for rapid assessment of contaminated sites and for screening purposes. However, because logistics demand that the equipment involved is portable, some of them are relatively expensive and do not always afford detection limits as low as those obtained by conventional laboratory instruments [15]. In addition, it is often necessary to "calibrate" or "train" the equipment with the analytes of interest. For example, electronic noses are specially trained through extensive chemometrics procedures [16], while PAS is calibrated with a particular VOC (e.g. toluene) and the other components of the air sample are determined as equivalents of that VOC [12, 13]. Measurements obtained in this way give little or no qualitative information about the constituent of the air sample. For example, Li et al. [17] measured the total volatile organic compounds (TVOCs) in an indoor microenvironment continuously with a photoacoustic Multi-Gas monitor but apart from formaldehyde, the other constituents of the samples were unknown.

Various types of portable gas chromatographs are now available for the direct measurements of VOCs. These include gas chromatographs with high-speed temperature and pressure programming and a GC ion mobility spectrometer [15]. In addition, portable GC time-of-flight (TOF) mass spectrometers [18] are available. However, GC/mass spectrometry (MS) is not routinely used for indoor air field measurements because of size, vacuum and energy requirements [15]. According to Santos and Galceran [15], portable gas chromato-graphs provide near real-time measurements, interactive sampling and quick solution to the problem faced at the time of the investigation. Nevertheless, they are usually expensive and are only able to achieve detection limits of the order of micrograms per cubic metre.

4.2
Sampling and Sample Analysis

Sampling can be done by passive or active techniques. Irrespective of the sampling technique adopted, subsequent laboratory analysis can be time-consuming and labour-intensive. Some of the common techniques used to collect and analyse indoor air samples are outlined in the following.

4.2.1
Active Air Sampling

This technique entails moving a predetermined volume of air at a controlled flow rate into a container or onto a sorbent. In its various forms, it is the most common technique used for the sampling of indoor VOCs.

4.2.1.1
Whole-Air Sampling

In whole-air sampling, a sufficient quantity of air is pumped into a container such as a polymer bag (Tedlar, Teflon or Mylar) [19] or a passivated stainless steel canister (e.g. SUMMA or silocan canisters) [20–23]. The attraction in using whole-air sampling is that sample collection is relatively simple and rapid, especially when time-weighted sampling is not required. In addition, the analyst has the opportunity to monitor the presence of a wide variety of polar and nonpolar VOCs from one sample and to carry out replicated analysis on the sample. Furthermore, there is no sample breakthrough (i.e. some of the analytes do not pass through the sampler without being held). However, loss of VOCs owing to chemical reactions within the container, physical adsorption by the walls of the container and dissolution in the water condensed in the container is not uncommon [24]. To minimise these causes, Tedlar bags should be protected from light by covering them with black bags and the internal surfaces of canisters should be electroplated or covered with siloxane [24]. Other short-comings associated with the use of this sampling method include the high cost involved in purchasing and the inconvenience in transporting canisters. Despite these drawbacks, it is the method of choice for sampling and storing very volatile hydrocarbons (e.g. C_2–C_4 compounds) and reactive compounds such as terpenes and aldehydes [25]. Hsieh et al. [24] recently showed that the half-lives of 56 VOCs, including several highly reactive alkenes in SUMMA canisters, Silocan canisters and Tedlar, were generally in excess of 30 days.

4.2.1.2
Sampling onto Sorbent Tubes

Excellent reviews on the use of sorbents for sampling air in general [14] and indoor VOCs in particular have appeared in the literature [5,11,26]. The most popular sorbents for sampling indoor VOCs can be classified into three broad categories: porous polymer-based sorbents (e.g. Tenax and Chromosorb), carbon-based sorbents (activated charcoal, graphitised carbon blacks, carbo-traps, anasorb, carboxens and carbosieve) and silica gels. Of these, porous polymers and carbon-based sorbents are the most widely used for indoor VOC sampling.

The choice of the sorbent material employed for a specific sampling depends on the absorption and desorption efficiencies of the sorbent for the target

VOCs as well as the stability of the VOCs on the sorbent. Additionally, the amount of VOCs retained on a sorbent is determined to a large extent by the sorbent bed length and the sorbent mass. Thus, a standard sorbent tube has a length of 16 cm, an outer diameter of 6 mm and contains 0.1–1 g of the sorbent(s) [27, 28]. Some parameters that should be considered when choosing the most appropriate sorbent method for a particular study include the "hydrophobicity", the "thermostability" and the "loadability" of the sorbent [5, 29]. The less water is retained by the sorbent, the less interference is experienced during analysis; the stabler the sorbent is, the more robust it is during thermal desorption of the analyte. Lastly, the more air that can be sampled onto a sorbent without sample breakthrough, the lower the detection limit that can be achieved.

Tenax TA, poly(2,6-diphenyl-p-phenylene oxide), is highly thermally stable and does not retain much water. In addition, it affords high desorption efficiency for a wide range of VOCs. Consequently, it is the most widely used sorbent for sampling multicomponent indoor VOCs in the carbon size range C_5–C_6 to C_{18}. The literature on indoor air is filled with examples of measurement studies conducted with this sorbent as the VOC trapping medium [2, 5, 28, 30–33]. However, care must exercised when using Tenax TA as a sorbent since it reacts with ozone and NO_x to form compounds which may facilitate the degradation of the sorbent [34]. To avoid this, ozone scrubbers must be used in conjunction with the sorbent, particularly when sampling is carried out in environments with high ozone concentrations [30].

When a single sorbent is not sufficiently efficient in capturing a wide range of VOCs, combinations of sorbents are employed to increase the range of compounds that can be confidently sampled. Consequently, multibed sorbents made up of Anasorb GCB1, Carbotrap and CarbopackB have been employed in some validated methods [30]. Similarly, multibed sorbents consisting of CarbopackC, CarbopackB and Carbosieve SIII [27]; and CarbopackB and Carbosieve SIII [10] have been used to trap a wide diversity of indoor VOCs.

Baltussen et al. [35] recently described the versatility of liquid poly(dimethylsiloxane) (PDMS) as a sorbent material for VOCs. Unlike other common solid sorbent materials, retention on PDMS occurs as a result of dissolution rather than adsorption. In addition, it also has several advantages over other forms of sorbents commonly used for indoor air sampling. For example, Baltussen et al. [35] showed that (1) it is more inert than other common sorbents and, therefore, it undergoes fewer reactions with the analytes and forms fewer artefacts, (2) it is more efficient in trapping polar compounds like organic acids and (3) it requires lower thermal desorption temperatures that other sorbents. Despite these advantages, PDMS is not as widely used in sorbent tubes for indoor VOC monitoring as Tenax; however, it is becoming more frequently employed in headspace sampling of VOCs, and as a fibre-coating material in solid-phase microextraction (SPME) [36, 37].

Active sampling onto sorbents entails storing known amounts of sorbent material(s) in glass or stainless steel tubes and drawing the sample through the tube by means of small battery-powered pumps. Since sorbents do not possess unlimited capacities to hold samples, caution must be exercised not to sample too much air onto the sorbent, otherwise "sample breakthrough" will occur. Representative samples are only obtained when the appropriate volume of air and a sorbent of the size that minimizes breakthrough are employed.

To minimise errors due to sample breakthrough, the total volume of the sample collected must be scrupulously monitored and a second bed of sorbent arranged in series with the first must be analysed. When the amount of a particular VOC in the second bed is greater than 5% of the amount in the first bed, sample breakthrough is implied [30]. While the safe sampling volumes (SSVs) suggested by US Environmental Protection Agency (EPA) method TO-17 for various VOCs is a useful sampling guide, care should be taken in applying the SSVs since breakthrough volumes are influenced by environmental factors like humidity. In keeping with this, US EPA method TO-17 [30] suggested that the sampling volume should not be greater than approximately 66% of the breakthrough volume [30].

Most classes of VOCs found in indoor environments are sampled onto sorbents by adsorption but highly reactive VOCs like carbonyl compounds are sampled by chemical reactions with the sorbent. Thus aldehydes and ketones are sampled by their reactions with sorbent gels coated with 2,4-dinitrophenylhydrazine to form stable hydrazones [38–40]. Similarly, formaldehyde has been sampled by its reaction with N-benzylethanolamine to give 3-benzyloxazolidine [41, 42].

Despite its widespread use in indoor VOC sampling, sorbent trapping provides no information about (1) all of the VOCs present in the sampled air since some VOCs are either not trapped by the sorbent(s) or are too reactive to remain on the sorbent surface and (2) the temporal variations in the concentrations of the VOCs that are being monitored.

4.2.2
Passive Air Sampling

4.2.2.1
Solid-Phase Microextraction

Koziel and Novak [37] recently reviewed the application of this combined sampling and sample preconcentration procedure to indoor air VOC measurement. Typically, a SPME sampler consists of a fused silica fibre that is coated by a suitable polymer (e.g. PDMS, PDMS/divinylbenzene, carboxen/PDMS) and housed inside a needle [37]. The fibre is exposed to indoor air and after sampling is complete, it is retracted into the needle until the sample is analysed. Compared with other sampling methods, it is simple to use and reasonably sensitive. However, samples collected by the procedure are markedly affected by environmental factors such as temperature. Therefore such samples cannot be stored for extended periods of time without refrigeration [36].

Koziel and Novak's [37] review of SPME is replete with examples of its use for (1) indoor VOC sampling followed by off-site laboratory analysis, (2) on-site sampling and analysis of indoor VOCs and (3) preconcentration of samples collected into canisters as well as (4) headspace sampling of the solvents extracted from samples collected by sorbent tubes. In addition to its ability to sample chlorinated VOCs [43], *n*-alkanes [44], aromatic hydrocarbons [45] and oxygenated hydrocarbons [46], the fibres of SPME can be doped with derivatising agents to make them amenable to sampling reactive VOCs like formaldehyde [47]. Despite its virtues, relatively few examples of the application of this technique for indoor air sampling have been described in the literature.

4.2.2.2
Passive Sampling onto Sorbents

The sorbents used for passive sampling are identical to those described for active sampling. The only difference is that while samples are pumped through the sorbents in the latter, they diffuse into the sorbents in the former. Woolfenden [48] has shown that the diffusive uptake rates of VOCs commonly found in indoor air on different sorbents vary from about 0.8 to 15 ng ppm^{-1} min^{-1}. Consequently, passive sampling is generally relatively slower than active sampling and may occur over several hours or days. Nevertheless, it is a popular sampling method, particularly for the evaluation of personal exposure. Brown et al. [49] recently used diffusive tubes packed with Tenax TA to monitor VOCs in 876 English homes, Schneider et al. [50] used it to measure indoor and outdoor levels of benzene, toluene and xylenes in German cities, while Son et al. [51] employed it for the measurement of VOCs in Korea. As in active sampling, chemical coated sorbents are also employed for the passive sampling of carbonyl compounds [47, 52].

4.2.3
Sample Desorption/Preconcentration

4.2.3.1
Whole Air Samples

Preconcentration of samples collected into canisters and polymeric bags is accomplished by passing known quantities of the samples through narrow capillary tubes held at very low temperatures by means of liquid cryogens [20, 53, 54]. The tubes are then rapidly heated to release the analytes into a cryofocussing unit and eventually to the gas chromatograph [19]. The procedure affords excellent recoveries for many VOCs but recoveries from samples stored in Tedlar bags are generally lower than those stored in canisters [24]. The main drawbacks of this procedure include the high cost of the cryogen and the susceptibility of the transfer tube to blockage.

4.2.3.2
SPME Samples

Extraction occurs when the needle of the syringe is exposed to fast moving, hot streams of gas within the injection port of the gas chromatograph [55].

4.2.3.3
Samples Collected onto Sorbents

Depending on the sorbents used, solvent desorption or thermal desorption may be applied to the sampled analytes. For silica gel and carbon-based sorbents, solvent desorption [40, 50, 51, 56, 57] and microwave desorption [5, 58] are the preconcentration methods of choice.

4.2.3.4
Solvent Desorption

Acetonitrile is frequently used for the desorption of 2,4-dinitrophenylhydrazones of carbonyl compounds collected on silica gel [39, 40, 59], while CS_2 is used for samples collected onto charcoal and dichloromethane for samples collected onto Anasorb 747 [59]. Carbon disulphide is particularly suitable for the desorption of nonpolar compounds but gives less satisfactory outcomes for the polar compounds. To overcome this shortcoming, polar cosolvents such as dimethylformamide, dimethylsulfoxide and ethanol are added to CS_2 to increase the recovery of polar analytes [36]. In addition, the use of CS_2 suffers from a number of other drawbacks, including the facts that (1) it reacts with amines and volatile chlorocarbons (2) it is unsuitable when electron detectors (e.g. electron capture detectors, ECDs) are used, (3) it is toxic and (4) has an unpleasant odour [36].

Compared with thermal desorption, solvent desorption is plagued by a number of shortcomings. For example, VVOCs are lost when the liquid sample is reconcentrated prior to its analysis. Moreover, solvent peaks may overlap with the peaks of VVOCs. A recent comparison of thermal and solvent desorption efficiencies also showed that with few exceptions solvent desorption consistently underestimates various classes of VOCs found in typical indoor air [59]. According to Wolkoff [5], solvent desorption leads to considerable loss in analytical sensitivity.

4.2.3.5
Thermal Desorption

This is a very popular method of transferring indoor VOC samples trapped by polymeric and carbon-based sorbents into analytical instruments. It usually entails running a stream of hot carrier gas (usually helium or argon) through the tubes in a direction opposite to that used for the sample collection. Typically, thermal desorption is carried out at around 250 °C [27]. After desorption, the compounds are reconcentrated by cryotrapping and then transferred directly by heat into the GC column. Although it affords greater sample desorption efficiency than solvent desorption, the desorbed sample can only be analysed once. Therefore, the only way to test the reproducibility of the method is to analyse multiple samples [36].

4.2.4
Characterisation of Indoor VOCs

Laboratory-based analyses of indoor VOCs are usually performed with gas chromatographs, which are coupled with flame-ionisation detectors (FIDs), ECDs or mass spectrometers. Alternatively high-performance liquid chromatography (HPLC) is used. Of these techniques, GC/MS provides the most conclusive qualitative and quantitative information, although a combination of a FID and an ECD has also been reported to permit the identification of compounds with widely different properties [12, 28]. Nonetheless, GC/MS remains the most widely used technique for the characterisation of indoor VOCs [15, 21, 51, 60, 61]. A total ion chromatogram is usually conducted to obtain global information on the ranges of compounds present and selected ion monitoring is performed to identify and monitor particular analytes. To facilitate the acquisition of quantitative information, the response factors of individual VOCs are often calculated against that of toluene, which is present in many indoor air samples. Typically, a splitless injection technique is employed [15, 31] to ensure the detection of compounds that are present at low levels. Various validated US EPA methods recommend the use of a dimethyl polysiloxane capillary column for the speciation and quantification of a suite of VOCs [22, 23, 30]. Similarly, the European Collaborative Action (ECA) report number 19 recommended a column with a polarity not exceeding that of 8% diphenyl polysiloxane [2]. Such columns are widely used in indoor air studies [10, 20, 21, 52, 62–64].

In order to increase the number of compounds that can be separated in a single analysis, it is not unusual to use a combination of GC columns with different polarities [28]. Temperature programming is also often required to achieve acceptable separation of analytes. A typical temperature program which has been used to separate different classes of indoor VOCs is summarised as follows: (1) hold at 40 °C for 1 min, (2) raise at 15 °C min^{-1} to 105 °C, (3) hold at 105 °C for 5 min, (4) raise at 20 °Cmin^{-1} to 245 °C and (5) hold at 245 °C for 5 min [10]. Column diameters ranging from 0.25 to 0.53 mm and lengths ranging from 25 to 100 m have been employed for indoor VOC measurements [10, 42, 51]. The choice of column dimensions depends on the properties of the compounds to be separated.

In their recent review of the application of GC in environmental analysis, Santos and Galceran [15] suggested that future perspectives of GC analysis include increasing use of

- GC/MS with positive and negative ion capabilities and sensitivities as low as parts per quadrillion.
- High-speed GC – with reduced sizes and capabilities for providing near real-time monitoring.
- Multidimensional GC – which remarkably increases the separation capabilities of the two columns used.
- GC-TOF-MS – with scanning capabilities of the order of 500 scans per second.

Such developments could impact future indoor VOC measurements markedly.

HPLC is mainly used for the analysis of the derivatives of low molecular weight carbonyl compounds such as formaldehyde [40, 59, 65]. However, formaldehyde is also quantified by a variety of other procedures, including the spectrometric acetyl-acetone method [66] and the chromotropic acid procedure [67].

4.2.5
Quality Assurance/Quality Control

Caution must be exercised to minimise errors at every stage of the characterisation. Therefore, quality assurance/quality control principles must be applied to sampling, sample storage, sample reconcentration and sample analysis.

4.2.5.1
Sampling

Short-term samplings are subject to temporal variations because of changes in source strength and ventilation conditions, while long-term measurements may show diurnal and seasonal variations [5]. These facts should be considered when planning sampling. Prior to sampling, sorbent tubes should be conditioned using a stream of carrier gas and temperatures that are higher than those that will be used for the analysis [27, 68]. Similarly, canisters need to be cleaned by repeated cycles of evacuation, flushing with humidified zero air and analysis for any trace levels of undesired gases [22, 24, 62]. As part of the quality control, sample breakthrough must be checked when sorbent tubes are used. In addition, field and method blanks as well as field duplicates must be collected and analysed [69].

4.2.5.2
Sample Storage

As a general rule, samples should be analysed as soon as possible after sampling and when immediate analysis is not feasible, they must be stored and transported under conditions that minimise artefact formation [34]. Thus samples trapped onto sorbent tubes are commonly covered with a Swagelok type of screw cap fitted with ferrules and stored in clean containers filled with nitrogen gas or activated charcoal [27, 48]. Tedlar bags must be protected from direct exposure to UV radiation, while canisters must be sealed airtight and transported to the laboratory in cool containers [24, 10].

4.2.5.3
Sample Desorption

Irrespective of the desorption method used (solvent or thermal) it is essential to ascertain the recovery efficiency of the VOCs of interest by spiking sorbent tubes and canisters.

4.2.5.4
Calibrations

Pumps should be calibrated with a rotameter [27] prior to and after sampling. Analytical instruments must also be calibrated before measurements. For example, GC/MS must be calibrated for mass and retention times using reference standard materials [70] and comparison made with the fragmentation patterns of known standards, usually a deuturated compound like toluene-d_8. Similarly, the method detection limit must be determined by finding the standard deviation of seven replicate analyses and multiplying it by the t-test value for 99% confidence of seven values [30, 62]. It is also usual for internal standards to be added to the samples and to evaluate the correlation coefficients of each standard used when multilevel calibration is employed. For automatic thermal desorption tubes, external and internal standardisations are achieved by injecting solutions of standards into the tubes [27, 28]; for canisters, solutions of standards are injected into the canisters followed by zero air.

In addition, the identity of each species must be obtained by comparing its retention time with that of an authentic standard sample or to an interlaboratory set of established retention times and by comparing its mass spectrum with that contained in a National Bureau of Standards or National Institute of Standards and Technology library installed on most modern instruments. It is usual to assign positive identification to a compound if its retention time is within 1% of that of the corresponding standard and the ratio of its quantifying ion to the target ion is not more that 10 times the standard deviation of the analogous ratio for an associated standard [33].

5
Current Knowledge on the Levels of VOCs in Indoor Microenvironments

Numerous studies have been undertaken to measure the level of indoor VOCs, in dwellings and offices in the past 10 years, and some results from such studies have been reviewed [5, 32]. A survey of two reference databases, CAPLUS and MEDLINE on SciFinder Scholar version 3 (American Chemical Society, 2001), confirmed that interest in the characterisation of VOCs in various environments has not abated. Although the survey did not capture all of the research done on indoor VOC, it gave a good indication of what has been done and where. One thousand four hundred and ninety three hits were recorded when "VOC, indoor air" was searched on SciFinder Scholar. Further searches conducted to find out where these studies were done revealed that most have been conducted in the USA and European countries, although some significant results have also emanated from Japan, Canada, Hong Kong, Australia, Brazil and China in the past 10 years. Representative examples of some of the studies are presented in Table 1, while the concentration levels of benzene, toluene,

Table 1 Selected indoor air volatile organic compound (*VOC*) studies conducted in the last decade

Type of indoor environment	Com-pounds	Country	Collection medium	Sample treatment	Analytical instrument	Reference
Office	11	Singapore	CarbopackB and Carbo-sieve SIII	Thermal desorption	GC-MS	[10]
Homes (kitchen cooking)	7	India	Tedlar bags	Cryogenically concentrated	GC-FID	[19]
Buildings (ten office, ten non-office)	43	China (Hong Kong)	Stainless canister	By TO-14 method	GC-MS	[21]
Homes	30	Finland	Tenax tubes	Thermal desorption (flash desorption)	GC-FID GC-MS	[31]
Library	5	Italy	ORBO-22 (formal-dehyde) CARBO-TRAP300 (VOC)	Isooctane desorption (formal-dehyde), thermal desorption (VOC)	GC-FID	[42]
Dwellings	33	Australia	Tenax TA, Ambersorb XE 340	Thermal desorption	GC-MS	[49]

Table 1 (continued)

Type of indoor environment	Com- pounds	Country	Collection medium	Sample treatment	Analytical instrument	Reference
Homes	6	Germany	OVM 3500 passive sampler badges	Carbon disulphide with internal standard (chemical desorption)	GC-FID	[50]
Homes	10	Korea	OVM 3500 passive sampler badges	Carbon disulphide (chemical desorption)	GC-MS	[51]
Homes	8	Holland	SKC charcoal tubes	Carbon disulphide (chemical desorption)	GC	[56]
Homes	50	Austria	Activated charcoal	Carbon disulphide	GC-MS	[61]
Homes	37	US	VOCCS, SUMMA canisters	By To-14 method	GC-MS	[63]
Homes	13	Canada	Tenax synthetic charcoal silanised glass beads	Thermal desorption	GC-FID; (TVOC) GC-MS	[64]
Office	12	Sweden	Sorbent (Tenax)	Not stated	GC-FID (TVOC), GC-MS (individual VOC), PAS (TVOC)	[71]
Office	60	Europe	Tenax Tubes	Thermal desorption	GC-FID; GC-MS PAS	[88]

Table 2 Arithmetic mean of the concentrations (μg m^{-3}) of benzene, toluene, ethylbenzene and xylenes in selected established buildings

Compound	Ref. [67]	Ref. [31]	Ref. [21]	Ref. [63]	Ref. [51]
	Australia	Finland	Hong Kong	USA	South Korea
Benzene	7.0	1.66	15.0	4.1	43.71
Toluene	14	5.62	206.3	15.3	170.67
m-Xylene, p-xylene	6.9	3.12	25.1	34.9	27.49[b]
Ethylbenzene	1.8	0.99	50.4	9.71	1.33
o-Xylene	8.9[a]	1.26	17.3	11.2	33.45

[a] o-Xylene/nonane.
[b] p-Xylene only.

ethylbenzene and xylenes in selected nonindustrial indoor air are presented in Table 2. The salient features of the studies reviewed reiterate some facts that were previously known, while others emphasise current trends.

1. GC/MS or GC/FID were used in most of the studies and are clearly the most popular detection methods used for VOC quantification. Nevertheless, for reactive carbonyl compounds such as aldehydes and ketones, HPLC analysis of their derivatised products is still the method of choice.
2. Compared with whole air sampling into Tedlar bags and canisters, active sampling onto sorbent materials is used more widely in these indoor air quality (IAQ) studies. Only a few studies made use of organic vapour monitor passive samplers. Of the sorbent materials used, Tenax is the most frequently employed, possibly because of its virtues, which are mentioned in Sect. 4.2.1. It has been used for the characterisation of aromatics, alkenes, cycloalkanes, aldehydes, ketones, esters, alcohols, terpenes, glycol derivatives and even amines [33, 59].
3. For samples collected onto sorbent materials, thermal desorption is used, except for a few instances where CS_2 desorption [50] and isooctane desorption [42] were preferred.
4. Generally, more studies have been conducted in residential indoor microenvironments than in offices. Although most of the studies were conducted in established rather than new buildings, many VOCs found in the former were also present in the latter but at higher concentrations. This is consistent with the thinking that VOCs emission rates from building materials decrease with the age of the building [67].
5. Only a few of the studies estimated the TVOC of the microenvironments reported [42, 67, 71] or focussed on complaint buildings [67, 72]. Because different definitions and methods of TVOC were employed to estimate the values quoted, it is difficult to make direct interstudy comparisons; nevertheless it appears that TVOC can range from 10 μg m^{-3} to several thousand micrograms per cubic metre in indoor environments.

Table 3 Frequency with which the 64 European Collaborative Action VOCs were monitored in selected studies

Always[a]	Frequently[b]	Normally[c]	Occasionally[d]	Never[e]
Benzene, toluene	Ethyl-benzene *o*-Xylene *m,p*-Xylene	1,2,4-Trimethyl-benzene, 1,3,5-trimethyl-benzene, styrene, *n*-nonane, *n*-decane, *n*-undecane, limonene	*n*-Propylbenzene, 2-ethyltoluene, naphthalene, 3-ethyltoluene, *n*-hexane, *n*-heptane, *n*-octane, *n*-dodecane, *n*-tridecane, *n*-pentadecane, *n*-hexadecane, 2-methylpentane, methylcyclopentane, 3-carene, *α*-pinene, *β*-pinene, 2-propanol, 1-butanol, 2-ethyl-1-hexanol, 2-ethoxyethanol, 2-butoxyethanol, butanal, hexanal, pentanal, nonanal, benzaldehyde, methylisobutylketone, acetone, trichloroethene, 1,1,1-trichloroethane, 1,4-dichlorobenzene, butylacetate, texanolisobutyrates	2-Pentafuran, tetrahydrofuran, isopropylacetate, 2-ethoxyethyl-acetate, ethylacetate, 4-phenylcyclo-hexene, 1-octene, 1-decene, 2-methoxy-2-propanol, 2-butoxyethoxy-ethanol, methylethyl-ketone, cyclohexanone, acetophenone

[a] Monitored in more than 75% of the studies.
[b] Monitored in 50–74% of the studies.
[c] Monitored in 25–49% of the studies.
[d] Monitored in 1–24% of the studies.
[e] None of the studies reported its concentration.

6. Direct comparison of the concentration levels found in each study is difficult since the sampling was conducted over different times, with different sampling techniques and different sample treatments and methods of analyses were used. However, the frequency with which the 64 VOCs that are of interest to the ECA are encountered in the different microenvironments can be classified as shown in Table 3.

The web-based Japanese automated formaldehyde data acquisition system/automated VOC data acquisition system (AFoDAS/AVODAS) [65], also showed that toluene was detected in 78% of the 1,422 homes monitored from 1998 to 2001 and that *p*-xylene, styrene, limonene and *α*-pinene were present in more than 50% of the homes. This data system corroborates the classification in Table 3 and highlights that since most of the VOCs are not frequently quantified, comparison between different studies is complicated.

5.1
The Total Volatile Organic Compounds Concept

Hundreds of VOCs are present in some typical indoor environments. It is therefore not practically possible to identify and quantify every compound, even with the most sensitive and selective techniques [5]. Consequently, different techniques have been used to express the TVOCs (for reviews of the methods see Refs. [2, 5, 12, 32]).

To redress problems caused by the different approaches to TVOC estimation, a uniform procedure was proposed [2, 12]. The procedure, which is based on sampling of VOCs on Tenax tubes followed by thermal desorption and GC/MS analysis (with nonpolar columns), proposed that TVOC be defined as

$$TVOC = S_{id} + S_{un},$$

where S_{id} is the sum of identified VOCs expressed in milligrams per cubic metre and S_{un} is the sum of unidentified VOCs relative to the response factor of toluene.

The procedure further recommends that as many VOCs as possible should be quantified in the analytical window bounded by the retention times of hexane and hexadecane, and that these VOCs should as far as possible include the 64 VOCs that are of special interest to the European Community [2]. A major shortcoming of the recommendation is that not all VOCs present in indoor air are included in the approach. For example, important indoor VOCs like 2-propanol, 2-methylpentane, 3-methylpentane and butanal elute before hexane while texanolisobutyrate elutes after hexadecane [60]. It was also expected that the definition would enhance interlaboratory TVOC values, classification and screening of indoor materials, and the identification of problems with ventilation design, indoor activities or materials [12]. However, De Bortoli et al. [73] observed large variances in interlaboratory studies performed with the approach. Nevertheless, it has been adopted in many recent indoor air studies [10, 60, 74].

6
Concepts for Regulating Indoor VOCs

Wolkoff [75] recently reviewed initiatives taken in Europe to reduce indoor air pollution by VOCs. Initiatives mentioned in the review include

- Source control.
- Control of emission from building materials.

- The establishment of a Europe-wide database of outdoor and indoor VOC levels through the EXPOLIS programme.
- IAQ audit projects.
- A labelling scheme.
- The establishment of guidelines for TVOC/VOCs.
- Avoiding nonessential VOCs.

In addition to these initiatives, various schemes aimed at reducing formaldehyde emission from building products, sensitising people to the effects of the presence of unsaturated fragrances in indoor air, use of labelled or low VOC, low isocyanate and acid anhydride emitting products have been introduced [75]. Because of their vast contribution to indoor VOCs, particularly in newly constructed buildings, a lot of the efforts just highlighted have focussed on the reduction of emissions from building products.

The US EPA [76] suggests, among other steps, that using household products as directed by the manufacturers and increasing ventilation when using household products could reduce indoor VOCs. In Japan, a database system for indoor formaldehyde and VOC (AFoDAS/AVODAS) has been established [65]. This should facilitate direct access to vital information needed by building designers, engineers and occupants to implement control measures [65]. In the USA, Hodgson et al. [39] have suggested the use of low VOC latex paints and carpet systems and decreased infiltration of unconditioned air. Mesaros [77] recently described the construction of "a low VOC house" in Australia, in which materials with low VOC emission factors like ceramics are used in preference to those with high VOC emission factors such as carpets. The house provides a good illustration of the use of source control in eliminating indoor VOCs. Another low VOC emission house, which employed low VOC emission materials and high ventilation rates, was independently reported by Guo et al. [78]. The marked difference in the TVOC levels in the house and those in normal houses provide support for the fact that IAQ can be improved through a combination of source control and building designs that minimise the negative impact of uncontrollable sources.

6.1
Source Apportionment

Unlike ambient VOCs, which originate predominantly from vehicular and industrial emissions, indoor VOCs have numerous and diverse origins. Therefore, source apportionment is an important factor in source control and the prime driving force for many IAQ studies. It can be accomplished by many methods, including the following.

6.1.1
Comparison of Indoor-to-Outdoor Concentration Ratios

This method assumes that the indoor-to-outdoor pollutant ratio depends on indoor and outdoor pollutant sources as well as the ventilation rates of the source and the sink, as shown in the following equation [79]:

$$C_I/C_0 = 1 + 1/C_0 \, (S_{source} - S_{sink})/(q_{source} - q_{sink}),$$

where q is the rate of ventilation, S_{sink} is indoor pollutant sinks, S_{source} is the indoor pollutant sources and C is the pollution concentration level.

When the indoor-to-outdoor pollutant ratio is approximately 1 for a VOC, it has comparable indoor and outdoor sources and when the ratio is greater than 1, it has dominantly indoor sources [38, 51, 52, 54, 80]. Typical indoor-to-outdoor pollutant ratio values for some VOCs are presented in Table 4 and these suggest that some VOCs, like toluene, have predominantly indoor sources, while others largely have outdoor sources. In the case of benzene, indoor air pollution is mainly caused by vehicle emissions from outside or by evaporation of gasoline from cars parked in underground car parks or in attached garages. Benzene as a constituent of commercial products has been banned in most European countries since the late 1970s or early 1980s and there are almost no more indoor sources of the compound. The method is often used in combination with statistical methods like Kruskal–Wallis, Wilcoxon W, and Kolmogorov–Smirnov Z tests [31, 54].

Table 4 Comparison of the ratios of the arithmetic mean of indoor concentration/arithmetic mean of outdoor concentration for selected VOCs in six homes. Data extracted from Lee et al. [52]

Compound	Living room	Kitchen
Benzene	2.35	2.05
Toluene	1.30	1.45
m-Xylene, p-xylene	0.62	0.84
o-Xylene	1.00	0.97
Ethylbenzene	0.54	0.75
1,3,5-Trimethylbenzene	0.82	0.77
Trichloroethane	1.06	1.24
Tetrachloroethene	0.52	0.48
1,4-Dichlorobenezene	0.90	1.03
Chloroform	2.00	2.38
Methylene chloride	0.92	0.84

Table 5 The use of factor analysis for source apportionment of VOCs identified in residential indoor air. Deduced from the data of Edwards et al. [81]

Percentage variance accounted for	Component	Associated VOC classes	Assigned source
18	1	Alcohols and alkanals	Cleaning products, fragrances, consumer products, particle board
18	2	n-Alkenes, substituted aromatics, hydrocarbons	Traffic emissions
17	3	Aromatics	Long-range transport
9	4	Alcohols and alkanals	Carpets, rubber, adhesives
6	6	Mainly 2-butoxyethanol	Cleaning products

6.1.2
Multivariate Data Analysis

These techniques reduce a large number of indoor VOCs to a few factors that can account for most of the cumulative variance in the VOC data [54, 81]. A factor loading matrix, which shows the correlation between the factors and the variables is often obtained. Edwards et al. [81] used this method to reduce 23 indoor VOCs in environmental tobacco smoke (ETS) free microenvironments to six factors and to apportion the most likely sources of the VOCs. A summary of the VOC classes loaded on each factor and the probable sources is presented in Table 5. It is, however, noteworthy that UNMIX and positive matrix factorisation, both of which are based on factor analysis and have been applied frequently to ambient air quality data [82], have not featured prominently in indoor VOC source apportionment reports.

6.1.3
Chemical Mass Balance Modelling

Watson et al. [83] recently undertook a review of the application of chemical mass balance modelling as a source apportionment technique for VOCs. The model assumes that the concentration of a chemical pollutant in a given sampling site is the summation of the contributions of all of the sources of the pollutants at the site. Thus, the concentration of the pollutant at the site can be predicted using the following equation [84]:

$$X_i = \sum_{j}^{p} a_{ij} S_j \quad i = 1, \ldots m,$$

where X_i is the predicted concentrations of pollutants at the site, a_{ij} is the source signature for pollutant i from source j, S_j is the contribution of source j, m is the number of pollutants (VOCs in this case) and p is the number of sources.

Although the model is frequently used to identify contributions of ambient VOCs from different sources [83], it has not been widely used for source apportionment of indoor VOCs. Won et al. [84], recently used the model to show that wall adhesive, caulking, I-beam joist and particleboard were the dominant sources of 24 indoor VOCs that were measured from a newly constructed building. However, similarities in the signatures of the various sources were observed. Such high correlations (collinearity) among measured chemical species could lead to large uncertainties in the estimated source contributions [83].

6.1.4
Instruments Used for Source Apportionment

The field and laboratory emission cell affords a portable, nondestructive method of testing the surfaces of potential VOC sources. In addition to its utility as a climatic chamber, it provides valuable information on source strength, which can be used for source apportionment and to formulate strategies for emission control. Wolkoff et al. [85] used it to identify emission processes in a number of building materials, while Jarnström and Saarela [66] recently utilised it to show that the dominant source of 2,2,4-trimethyl-1,3-pentadiol-diisobutyrate in the indoor air of some problem apartments was the floor surface.

Apart from the field and laboratory emission cell, other instruments that have been used for source identification and apportionment include

- Direct measurements by portable instruments [71].
- Passive samplers [86].
- Headspace samplers [87].
- Multisorbent tubes [68].

6.2
Understanding Emissions from Indoor Sources

Indoor VOC levels are influenced by a large number of factors [5, 10, 88]. The most prominent ones include (1) air exchange rate, (2) source characteristics, (3) ventilation systems, (4) meteorology (temperature and relative humidity), (5) age of a building, (6) building design, (7) type of indoor activities (e.g. cooking, smoking and photocopying), (8) sorption, desorption and deposition rates, (9) mixing and distribution of pollutants and (10) removal rate.

Of these factors, source characteristics, particularly characteristics of building materials, have been the most explored in the literature. Thus various studies have attempted to link emission rates and sink effects of building and furnishing materials with indoor VOC levels [84, 89–91]. Many studies indicate that emission levels in new buildings are much higher than those in established buildings [39, 67, 92]. This is possibly because emissions from building materials generally exhibit a decaying profile that is illustrated by the following equation [67, 72]:

$$EF = M_0 k_1 \exp(k_1 t),$$

where EF is the emission factor for the source material (usually expressed in micrograms per square metre per hour), M_0 is the quantity of pollutant on the surface of the material (usually expressed as micrograms per square metre) and k_1 is the decay constant (per hour).

Alternatively, when there are multiple decay processes, the following equation is useful:

$$EF = EF_{01} \exp(-k_1 t) + EF_{02} \exp(-k_2 t),$$

where EF_{01} and EF_{02} are the initial decay constants for two simultaneous decay processes. Thus, the decay is initially fast and the VOC level is higher in new buildings [67] as illustrated by Fig. 1.

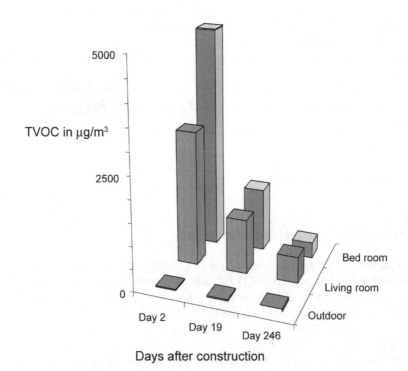

Fig. 1 Dependence of total volatile organic compounds (*TVOC*) on the age of a newly constructed home. Data from Ref. [67]

Another approach that is commonly used to evaluate the emission rates of VOCs in indoor microenvironments is to estimate an area-specific emission rate (SER). This approach assumes that VOCs are homogeneously mixed in the environment and that SER can be calculated with the following equation [10]:

$$SER = NV (C_1 - C_0)/A,$$

where the SER is in micrograms per square metre per hour, V is the volume of the space (cubic metres), N is the air exchange or infiltration rate (per hour), A is the floor area space (square metres), C_1 is the indoor concentration (micrograms per cubic metre) and C_2 is the outdoor concentration (micrograms per cubic metre)

Van Winkle and Scheff [63] used a variant of this equation to show that indoor VOCs have predominantly indoor sources, while Hodgson et al. [39] reported the SER values for a wide range of VOCs in "manufactured" and "site-built" homes in the USA. As expected, many of the VOCs monitored by Hodgson et al. [39] showed decreased emission rates with the age of the building.

It is noteworthy that the VOC emission rates of building materials vary widely [67, 73, 93–95]. Thus, Mølhave [94] reported over 20 years ago that the emission rates of wall/flooring glue, water-based poly(vinyl alcohol) glue and gypsum board are of the order 2.7×10^5, 2.1×10^3 and 30 $\mu gm^{-2}h^{-1}$ respectively. More recently, Kwok et al. [95] reported that VOC emission rates for varnish-painted aluminium, plaster, gypsum and plywood for toluene, o-xylene, m-xylene, p-xylene and ethylbenzene are also significantly different; with that of aluminium being approximately 65% higher than that of plywood as illustrated by Fig. 2.

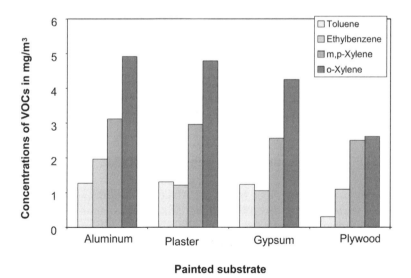

Fig. 2 Effect of substrate on volatile organic compound (*VOC*) emissions from indoor building materials. Constructed from the data of Kwok et al. [95]

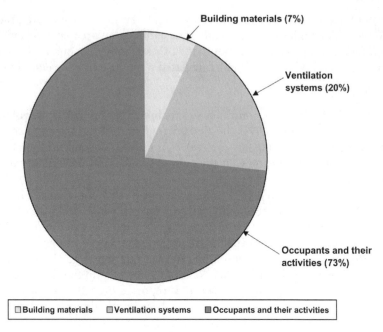

Fig. 3 Source apportionment of indoor aromatics. Data from Zuraimi et al. [68]

It is also worth mentioning that emission rates vary with different indoor activities. Van Winkle and Scheff [63] associated 1,1,1-trichloroethane emission factors of 353, 522, 988, 1,419 and 2,790 µg h^{-1} with the presence of a washer/drier in a utility room, storage of hair products, storage of chemicals, periodic dry cleaning and storage of mothballs, respectively. Occupants of air-conditioned offices and their activities have also been shown to contribute more to VOCs levels in commercial offices in Singapore than ventilation systems and building materials [10], as illustrated by Fig. 3. This result corroborates the findings of Hodgson et al. [96], which suggested that occupants of new office buildings contributed more VOCs to the indoor air than other sources.

6.3
Understanding the Interaction of VOCs with Indoor Materials

Diffusive interactions of VOCs with the surfaces of building walls, floors and household materials have been the subject of several experimental investigations, modelling and simulation [90, 97–100]. Such interactions regulate peak levels of indoor VOCs, while subsequent desorption of the adsorbed VOCs delays their disappearance from indoor environments. In order to predict and model the VOC emission rates of indoor materials, it is essential to know the diffusion and partition coefficients of individual VOCs. Bodalal et al. [98] recently described an experimental method for the determination of the diffusion and partition coefficients of some VOCs. This method should enhance the prediction of VOC emissions from indoor materials without recourse to elaborate chamber tests.

Secondly, VOCs react with indoor ozone to produce sub-micron-sized particles [101, 102]. Thus, terpenes, which are commonly found in many household consumer products, interact with ozone, which is also widespread in indoor air through outdoor infiltration and the use of office equipment like laser printers and photocopiers, to form particles. Such reactions can markedly increase the number concentrations and mass concentrations of sub-micron-sized particles. In addition, styrene and 4-phenylcyclohexene react with ozone to generate appreciable amounts of aldehydes [4, 101].

6.4
Indoor VOC Guidelines

According to Mølhave [103], "A guideline is a set of criteria (i.e. standards for making judgements) specifically assembled to indicate threshold levels of a harmful or no noxious agent consistent with the good health." The first notable attempt to provide some guidelines for indoor VOCs was made by Seifert [104], in which he classified indoor VOCs into alkenes, aromatic hydrocarbons, terpenes, halocarbons, esters, aldehydes and ketones (excluding formaldehyde) and proposed that

- TVOC should not exceed 300 µg m^{-3}.
- No individual compound should have a concentration greater than 10% of TVOC or 10% of the concentration apportioned to that class of VOC.

Pluschke [105] recently reviewed Seifert's paper [104] and showed that only a few countries have guidelines for indoor TVOC. The USA has a value of 200 µg m^{-3} [106], Germany 300 µg m^{-3} [104], Australia 500 µg m^{-3} [107] and Finland has various values ranging from 200 to 600 µg m^{-3} [108]. Pluschke [105] also stated that Seifert modified his original concept to include a target value defined as 200–300 µg m^{-3} and a recommendation for official intervention if the TVOC concentration exceeds 1,000–3,000 µg m^{-3}.

Guidelines for individual VOCs are also available in some countries. In Poland, the maximum allowable concentrations for some VOCs has been set at 10 µg m^{-3} for benzene, 200 µg m^{-3} for toluene, 100 µg m^{-3} for butyl acetate, 100 µg m^{-3} for ethylbenzene, 100 µg m^{-3} for m-xylene, 20 µg m^{-3} for styrene and 30 µg m^{-3} for p-diclorobenzene [109]. In the context of the 64 VOCs of interest to the European Commission (ECA-IAQ 1995) only toluene, 2-ethoxyethanol, 2-butoxyethoxyethanol and 1-methoxy-2-propanol have readily available guideline values [110]. The odour threshold, sensory irritation exposure limit and health-based indoor exposure limits of these VOCs are presented in Table 6.

Despite the ubiquitous nature and importance of VOCs in indoor environments it is surprising that no international indoor VOC guideline has emerged.

Table 6 Guideline levels for some VOCs (according to Refs. [105, 110])

Compound	Odour threshold $(mg\ m^{-3})$	Sensory irritation exposure limit estimate $(mg\ m^{-3})$	Health-based indoor air exposure limit estimate $(mg\ m^{-3})$
Toluene	1	8	8
2-Ethoxyethanol	4.6	10	0.4
2-Butoxyethoxyethanol	0.0092	–	9
1-Methoxy-2-propanol	0.7	10	10

Nielson et al. [110] linked the difficulty experienced in evaluating sensory and health effects of indoor VOCs with the absence of indoor VOC standards and guidelines. Although European Commission report number 19 [2] recommended that indoor VOCs should be kept as low as reasonably achievable, more concerted efforts should be made to formulate universally acceptable sets of guidelines for as many indoor VOCs as possible.

7
Health Effects of Indoor VOCs

Mølhave [103] suggested that the health effects of VOCs could be grouped into

- Immune effects and other hypersensitivity effects (e.g. asthma and allergy).
- Cellular effects (e.g. cancer).
- Cardiovascular effects.
- Neurogenic and sensory effects (e.g. odour and irritation).
- Respiratory effects other than immunological.

The US EPA website [76], accessed in September 2003, provides more details, suggesting that the health effects of indoor VOCs include eye, nose and throat irritations, headaches, loss of coordination, nausea, damage to the liver, kidneys, and the central nervous system, and cancer. Similarly, health effects of ETs, which contains thousands of VOCs as well as nicotine, 3-ethenylpyridine, carbon monoxide and particulate matter, include eye, nose and throat irritation, carcinogenic effects, activation of the immune system, exacerbation of asthma and respiratory tract illnesses [6].

While individual VOCs like benzene and toluene have been linked with acute myeloid leukaemia and neurotoxicity, respectively [111, 112], epidemiological studies of the health effects of indoor VOCs have related TVOC rather than individual VOC levels to exposure. The outcomes of such studies have been mixed [113]. In some cases, positive associations between SBS, building-related illness or multiple chemical sensitivity syndromes and TVOC levels were observed

[114–116], while negative associations were reported by Sundell et al. [117]. Yet, no association was found in a few studies [118].

Several reasons may be adduced for the inconsistent association between SBS complaints and TVOC levels [5, 12]. Some of these are outlined in the following:

- Indoor air chemistry leads to the formation of VOCs and other species that are different from those monitored. For example, ozone reacts with VOCs to give secondary products that could be responsible for the observed SBS [102].
- Only compounds in a narrow chromatographic window are normally monitored; low molecular weight aldehydes, which may play a significant role in SBS, are not routinely monitored as part of TVOC [119].
- Ventilation systems are significantly associated with SBS complaints [120, 121].
- Particles present in an indoor environment might contribute significantly to SBS systems [5].
- ETS is associated with many SBS-type symptoms [5].
- Measurements are usually carried out as mean time-integrated concentrations at the centre of the room rather than in the breathing zones of the subject [5].
- Self-reporting questionnaires are subjective means of assessing SBS [115].
- The influences of psychosocial factors are being ignored [116].
- TVOC is not biologically important [119].
- Biologically important VOCs have not been found and are not being monitored [119].

Thus the role of VOCs in SBS complaints is far from being understood. More research is particularly required in the

- Development of validated methods for TVOC and the dose–response relationship.
- Risk indicators for multiple exposures [113].
- Evaluation of TVOC and SBS/health effects from carefully designed epidemiological studies.
- Development of universal guidelines for evaluating exposures [122].

Since TVOC does not permit a consistent dose–response relationship, Mølhave [102] suggested that it should be treated as an indicator of the presence of VOCs and used only for source identification, IAQ assessment and as a screening tool for exposure assessment rather than a guideline or an official recommendation.

However, despite the statistically insignificant difference between the TVOC values in buildings with and without SBS problems, the underlying difference among such buildings was manifested in Cooman's plot and partial least squares discriminate analysis plots [123]. Similarly, principal component analysis has been used to separate buildings with low and high prevalence of SBS [124]. It therefore appears that multivariate projection methods could play significant roles in the identification of causality in indoor VOC exposure studies.

8
Trends/Perspectives

It is evident from the previous sections that

- The TVOC concept has limited use and must be used with caution [119].
- A compound-to-compound approach of evaluating the health effects of organic compounds in indoor air as suggested by Wolkoff [75] should be explored alongside the TVOC approach.
- More research is required on the role of reactive chemistry in the causation of SBS.
- The roles of ionic species, hydroxyl and peroxide radicals as well as substances absorbed onto particles in the causation of SBS should be researched [125].
- Attempts have been made to estimate OH radicals through modelling and indirect measurements [125], but no direct measurements have been made to date. Future work is required in this area.
- More research is required in order to understand the health effects of secondary products like ketones, peroxy-acetyl nitrate and organic acids, which are generated from the reactions of VOCs in indoor air [125].
- More indoor air audits are required in developing nations.
- A collaborative approach from environmental scientists and health agencies is required.
- Practical applications of the knowledge gained in the various areas of IAQ are desirable.

9
Concluding Remarks

The past decade has witnessed tremendous growth in indoor air audits, development of a quasi-uniform definition of VOC levels in indoor micro-environments and improvement of concepts for screening IAQ and assessing exposure. Significant advancements have been made in emission modelling, source identification, source control, source characteristics and interaction of indoor VOCs with indoor materials. Some progress, albeit slow, has occurred in the development of universally acceptable exposure guidelines. The introduction of smaller, faster and smarter instrumentation to the market could enhance fieldwork markedly. However, the link between health/sensory effects and indoor VOCs levels is still largely unclear. Further work is urgently required in this area and in the search for insight into the role of reactive chemistry in the generation, degradation and transformation of indoor VOCs.

Acknowledgements I would like to thank Lidia Morawska for her encouragement, Erik Uhde for useful comments and James Blinco for typing part of the manuscript and the tables.

References

1. World Health Organisation (1989) Indoor air quality: organic pollutants. EURO reports and studies no 111. World Health Organisation, Copenhagen, p 1
2. European Collaborative Action on Indoor Air Quality and its Impact on Man (1997) Total volatile organic compounds (TVOC) in indoor air quality investigations. Report no 19, EUR 17675EN. Luxembourg Office for Official Publications of the European Community
3. Singer BC, Hodgson AT, Guevarra KS, Hawley EL, Nazaroff WW (2002) Environ Sci Technol 36:846
4. Morrison GC, Nazaroff WW (2002) Environ Sci Technol 36:2185
5. Wolkoff P (1995) Indoor Air Supp 3:9
6. Jones AP (1999) Atmos Environ 33:4535
7. Bluyssen PM, De Oliveira Fernandes E, Groes L, Clausen G, Fanger PO, Valbjørn, O, Bernhard CA, Roulet CA, (1996) Indoor Air 6:221
8. Salthammer T (ed) (1999) Organic indoor air pollutants. Wiley-VCH, Weinheim
9. Wilkins K (2002) Microbial VOC (MVOC) in buildings, their properties and potential use. Proceedings of the 9th international conference on indoor air quality and climate, Monterey, CA, USA, p 431
10. Zuraimi MS, Tham KW, Sekhar SC (2003) Build Environ 38:23
11. ECA (1994). Sampling strategies for volatile organic compounds (VOCs) in indoor air. Commission of the European Communities, Brussels, Luxemburg
12. Mølhave L, Clausen G, Berglund B, De Ceaurriz J, Kettrup A, Lindvall T, Maroni M, Pickering AC, Risse U, Rothweiler H, Seifert B, Younes M (1997) Indoor Air 7:225
13. Krueger U, Kraenzmer M, Strindehag O (1995) Environ Int 21:791
14. Ekberg L (1999) In: Salthammer T (ed) Organic indoor air pollutants. Wiley-VCH, Weinheim, p 73
15. Santos FJ, Galceran MT (2002) Trends Anal Chem 21:672
16. Morvan M, Talou T, Beziau J-F (2003) Sens Actuators B 95:212
17. Li YY, Wu P-C, Su H-J, Chou P-C, Chiang CM (2002) Effects of HVAC ventilation efficiency on the concentrations of formaldehyde and total volatile organic compounds in office buildings. Proceedings of the 9th international conference on indoor air quality and climate. Monterey, CA, USA, p 376
18. Syage JA, Nies BJ, Evan MD, Hanold KA (2001) J Am Soc Mass Spectrom 12:648
19. Pandit GG, Srivastava PK, Mohan Rao AM (2001) Sci Total Environ 279:159
20. Lee SC, Li WM, Chan LY (2001) Sci Total Environ 279:181
21. Chao CY, Chan GY (2001) Atmos Environ 35:5895
22. US Environmental Protection Agency (1988), Compendium method TO-14: determination of volatile organic compounds (VOCs) in ambient air using SUMMA passivated canister sampling and gas chromatography analysis. Environmental Monitoring Systems Laboratory, US Environmental Protection Agency, Research Triangle Park, NC
23. US Environmental Protection Agency (1997), Compendium method TO-15: determination of volatile organic compounds (VOCs) in air collected in specially-prepared canisters and analyses by gas chromatography/mass spectrometry (GC/MS). US Environmental Protection Agency, Cincinnati, OH, EPA 625/R-96/010b
24. Hsieh C-C, Horng S-H, Liao P-N (2003) Aerosol Air Qual Res 3:17

25. Batterman SA, Zhang G-Z, Baumann (1998) Atmos Environ 32:1647
26. Uhde E (1999) In: Salthammer T (ed) Organic indoor air pollutants. Wiley-VCH, Weinheim, p 4
27. Wu C-H, Lin M-N, Feng C-T, Yang K-L, Lo Y-S, Lo J-G (2002) J Chromatogr A 996
28. Mattinen ML, Tuominen J, Saarela K (1995) Indoor Air 5:56
29. Camel V, Caude M (1995) J Chromatogr A 170:3
30. US Environmental Protection Agency (1999) Compendium of methods for the determination of toxic organic compounds in ambient air. Compendium method TO-17: determination of volatile organic compounds (VOCs) in ambient air using active sampling onto sorbent tubes. US Environmental Protection Agency, Cincinnati, OH
31. Edwards RD, Jurvelin J, Saarela K, Jantunen M (2001) Atmos Environ 35:4531
32. Brown SK, Sim MR, Abramson MJ, Gray CN (1994) Indoor Air 4:123
33. Zhu J, Aikawa B (2003) Environ Int (in press)
34. Cao X-L, Hewitt CN (1994) J Chromatogr A 688:368
35. Baltussen E, David F, Sandra P, Janssen H-G, Cramers CA (1998) J High Resolut Chromatogr 21:332
36. Harper M (2000) J Chromatogr A 885:129
37. Koziel JA, Novak I (2002) Trends Anal Chem 21:840
38. Brickus LSR, Cardoso JN, De Aquino Neto FR (1998) Environ Sci Technol 32:3485
39. Hodgson AT, Rudd AF, Beal D, Chandra S (2000) Indoor Air 10:178
40. US Environmental Protection Agency (1999) Compendium method TO-11A: determination of formaldehyde in ambient air using adsorbent cartridges followed by high performance liquid chromatography (HPLC) (active sampling methodology). US Environmental Protection Agency, Cincinnati, OH
41. Seppänen OA, Fisk WJ, Mendell MJ (1999) Indoor Air 9:226
42. Righi E, Aggazzotti G, Fantuzzi G, Ciccarese V, Predieri G (2002) Sci Total Environ 286:41
43. Chai M, Arthur CL, Pawlizyn J, Belardi RP, Pratt KF (1993) Analyst 118:1501
44. Koziel JA, Jin M, Khaled A, Naoh J, Pawliszyn J (1999) Anal Chim Acta 400:153
45. Elke K, Jermann E, Begerow J, Dunemann L (1998) J Chromatogr A 826:191
46. Saba A, Raffaelli A, Pucci S, Salvadori P (1999) Rapid Commun Mass Spectrom 13:1899
47. Koziel JA, Naoh J, Pawliszyn J (2001) Environ Sci Technol 35:1481
48. Woolfenden EA (1995) In: Subramanian G (ed) Quality assurance in environmental monitoring instrumental methods. VCH, Weinheim, p 133
49. Brown VM, Coward SKD, Crump DR, Llwellyn JW, Mann HS, Raw GJ (2002) Indoor air quality in English homes – VOCs. Proceedings of the 9th international conference on indoor air quality and climate, Monterey, CA, USA, p 477
50. Schneider P, Gebefugi I, Richter K, Wolke G, Schnelle J, Wichmann HE, Heinrich J (2001) Sci Total Environ 267:41
51. Son B, Breysse P, Yang W (2003) Environ Int 29:79
52. Lee SC, Li WM, Ao CH (2002) Atmos Environ 36:225
53. Li WM, Lee SC, Chan LY (2001) Sci Total Environ 273:27
54. Guo H, Lee SC, Li WM, Cao JJ (2003) Atmos Environ 37:73
55. Martos PA, Saraullo A, Pawliszyn J (1997) Anal Chem 69:402
56. Fischer PH, Hoek G, van Reeuwijk H, Briggs DJ, Lebert E, van Wijnen JH, Kingham S, Elliott PE (2000) Atmos Environ 34:3713
57. Otson R, Fellin P, Tran Q (1994) Atmos Environ 28:3563
58. Vu Duc T, Huynh CK (1991) J Chromatogr Sci 29:179
59. Zorn C, Köhler M, Weis N (2002) Underestimation of indoor air concentrations; comparison of Tenax-adsorption/thermal desorption with different sorbent/liquid extraction. Proceedings of the 9th international conference on indoor air quality and climate, Monterey, CA, USA, p 595

60. Massold E, Riemann A, Salthammer T, Schwampe W, Uhde E, Wensing M, Kephalopoulos S (2000) Determination of response factors for GC/MS of 64 volatile organic compounds for measuring TVOC. Proceedings of healthy building 2000 vol 4. Helsinki, Finland, p 91

61. Hutter HP, Moshammer H, Wallner P, Damberger B, Tappler P, Kundi M (2002) Volatile organic compounds and formaldehyde in bedrooms: results of a survey in Vienna, Austria. Proceedings of the 9th international conference on indoor air quality and climate. Monterey, CA, USA, p 239

62. Lee SC, Guo H, Li WM (2002) Atmos Environ 36:1929

63. Van Winkle MR, Scheff PA (2001) Indoor Air 11:49

64. Shaw CY, Salares V, Magee RJ, Kanabus-Kaminska M(1999) Build Environ 34:57

65. Park JS, Ikeda K (2003) Indoor Air 13:35

66. Jarnström H, Saarela K (2002) The apportionment of volatile organic compounds and the effect of remedial actions at the indoor air sites. Proceedings of the 9th international conference on indoor air quality and climate. Monterey, CA, USA, p 625

67. Brown SK (2002) Indoor Air 12:55

68. Zuraimi MS, Tham KW, Sekhar SC (2002) Identification and quantification of VOCs sources in air-conditioned office buildings in Singapore. Proceedings of the 9th international conference on indoor air quality and climate. Monterey, CA, USA, p 183

69. Adgate JL, Bollenbeck M, Eberly LE, Stroebel C, Pellizzari ED, Sexton K (2002) Residential VOC concentrations in probability based sample of household with children. Proceedings of the 9th international conference on indoor air quality and climate. Monterey, CA, USA, p 203

70. Shields HC, Fleischer DM, Weschler CJ (1996) Indoor Air 6:2

71. Ekberg LE (1994) Atmos Environ 28:3571

72. Brown SK (1999) Indoor Air 9:209

73. DeBortoli M, Kephalopulous S, Kirchner S, Schauenburg H, Vissers H (1999) Indoor Air 9:103

74. Massold E, Riemann A, Salthammer T, Schwampe W, Uhde E, Wensing M, Kephalopoulos S (2000) Comparison of TVOC by GC/MS with direct reading instruments. Proceedings of healthy building 2000, vol 4. Helsinki, Finland, p 67

75. Wolkoff P (2003) Indoor Air 13:5

76. http://epa.gov/iaq/woc.html

77. Mesaros D (2003) Clean Air J 37:29

78. Guo H, Murray F, Lee SC (2003) Build Environ 38:1413

79. Chan AT (2002) Atmos Environ 36:1543

80. Daisey JM, Hodgson AT, Fisk WJ, Mendell MJ, Brinke JT (1994) Atmos Environ 28:3557

81. Edwards RD, Jurvelin J, Koistinen K, Saarela K, Jantunen M (2001) Atmos Environ 35:4829

82. Hopke PF (2003) J Chemometrics 17:255

83. Watson JG, Chow JC, Fujita EM (2001) Atmos Environ 35:1567

84. Won D, Shaw CY, Biesenthal TA, Lusztyk E, Magee RJ (2002) Applications of chemical mass balance modeling to indoor VOCs. Proceedings of the 9th international conference on indoor air quality and climate. Monterey, CA, USA, p 268

85. Wolkoff P, Clausen PA, Nielson PA, Gunnarsen K (1993) Indoor Air 3:291

86. Seifert B, Ullrich D (1987) Atmos Environ 21:195

87. Yoshizawa S, Ezoe Y, Goto S, Maeda T, Endo O, Watanabe I (2002) A simple method to determine the sources of VOCs in indoor air. Proceedings of the 9th international conference on indoor air quality and climate. Monterey, CA, USA, p 938

88. Bluyssen PM, Cox C, Sappänen O, de Oliveira Fernandes E, Clausen G, Müller B, Roulet C-A (2003) Build Environ 38:209

89. European Collaborative Action on Indoor Air Quality and its Impact on Man (1997) Evaluation of emission from building products: solid flooring materials, report no 18. EUR 17334EN. Luxembourg Office for Official Publications of the European Community

90. Zhang LZ, Niu JL (2003) Build Environ 38:939

91. Afshari A, Lundgren B, Ekberg LE (2003) Indoor Air 13:156

92. Brown SK (2000) Air toxics in a new Victorian dwelling over an eight-month period. Proceedings of the 15th international clean air and environment conference, 26–30 November 2000, Melbourne, vol 1, p 458

93. VOCEM (1998) Further development and validation test chamber method for measuring VOC emissions from building materials and products. Champs sur Marne (Marne la Vallee, VOCEM) (SMT4-VT95-2039 interlaboratory comparison report)

94. Mølhave L(1982) Environ Int 8:117

95. Kwok NH, Lee SC, Guo H, Hung W-T (2003) Build Environ 38:1019

96. Hodgson AT, Daisey JM, Grot RA (1991) J Air Waste Manage Assoc 41:1461

97. Meininghaus R, Salthammer T, Knöppel H (1999) Atmos Environ 33:2395

98. Bodalal A, Zhang JS, Plett EG (2000) Build Environ 35:101

99. Meininghaus R, Udhe E (2002) Indoor Air 12:215

100. Xu Y, Zhang Y(2003) Atmos Environ 37:2497

101. Weschler CJ, Shields HC (1999) Atmos Environ 33:2301

102. Fielder N, Zhang J, Fan Z, Kelly-McNeil K, Lioy P, Gardner C, Ohman-Strickland P, Kipen H (2002) Health effects of a volatile organic mixture with and without ozone Proceedings of the 9th international conference on indoor air quality and climate. Monterey, CA, USA, p 596

103. Mølhave L (2003) Indoor Air 13:12

104. Seifert B (1990) In: Walkinshaw DS (ed), Indoor air '90. Proceedings of the 5th international conference on indoor air quality and climate. July 29–August 5, Toronto, Canada, p 35

105. Pluschke P (1999) In: Salthammer T (ed) Organic indoor air pollutants. Wiley-VCH, Weinheim, p 291

106. USA-EPA (1996) Indoor air quality updates 9 and 10:8

107. NHMRC (1993) Volatile organic compounds in indoor air. Report of 115th session, Canberra, Australia

108. FISIAQ/Finnish Society of Indoor Air Quality and Climate, Finnish Association of Construction Clients/RAKI Finnish Association of Architects/SAFA, Finnish Association of Consulting Firms/SKOL (1995) Classifications of indoor climate, construction and furnishing materials, Helsinki

109. Regulation of the Minister of Health and Social Security of Poland from March 12, 1996 on maximum allowable concentrations of pollutants emitted by building materials and furnishings in inhibited closed areas

110. Nielsen GD, Hansen LF, Nexø BA, Poulsen OM (1998) Indoor Air Supp 5:37

111. Dor F, Dab W, Empereur-Bissonnell P, Smirdu D (1999) CRT Rev Toxicol 29:129

112. Liu Y, Fetcher L (1997) ToxicolAppl Pharmacol 142:270

113. Andersson K, Bakke JV, Bjorseth O, Bornehag C-G, Clausen G, Hongslo JK, Kjellman M, Kjaergaard S, Levy F, Molhave L, Skerfving S, Sundell J (1997) Indoor Air 7:78

114. Norback D, Bjornsson E, Janson C, Widstrom J, Boman G (1995) Occup Environ Med 52:388

115. Yoshino H, Amano K, Ikeda K, Nozaki A, Iida N, Katuta K, Hojo S, Ishikawa S (2002) Field survey on indoor air quality and occupants' health conditions in sick houses. Proceedings of the 9th international conference on indoor air quality and climate. Monterey, CA,USA, p 119

116. Herzog V, Witthauer J, Brasche S, Bischof W (2002) VOC-related sensory symptoms in office buildings. Proceedings of the 9th international conference on indoor air quality and climate. Monterey, CA, USA, p 78
117. Sundell J, Anderson B, Anderson K, Lindvall T (1993) Indoor Air 3:82
118. Nielsen GD, Alarie Y, Poulsen OM, Nexø BA (1995) Scand J Work Environ Health 21:165
119. Wolkoff P, Nielsen GD (2001) Atmos Environ 35:4407
120. Seppänen O, Fisk WJ (2002) Indoor Air 12:98
121. Daisey JM, Angell WJ, Apte MG (2003) Indoor Air 13: 53
122. Mølhave L (1998) Indoor Air (Suppl 4):17
123. Sunesson A-L, Gullberg J, Olsson-Kohler K, Blomquist G (2002) Multivariate evaluation of VOCs in homes and office buildings with and without known SBS complaints Proceedings of the 9th international conference on indoor air quality and climate. Monterey, CA, USA, p 84
124. Pommer L, Fick J, Andersson B, Sundell J, Nilsson C, Sjostrom M, Stenberg B (2002) Class separation of buildings with high and low prevalence of SBS principal component analysis. Proceedings of the 9th international conference on indoor air quality and climate. Monterey, CA, USA, IV, p 96
125. Carslaw N (2003) Atmos Environ 37:5645

The Handbook of Environmental Chemistry Vol. 4, Part F (2004): 37–71
DOI 10.1007/b94830
© Springer-Verlag Berlin Heidelberg 2004

Emissions of Volatile Organic Compounds from Products and Materials in Indoor Environments

Tunga Salthammer (✉)

Fraunhofer Wilhelm-Klauditz-Institut (WKI), Material Analysis and Indoor Chemistry,
Bienroder Weg 54 E, 38108 Braunschweig, Germany
salthammer@wki.fhg.de

Abstract Building products, furnishings and other indoor materials often emit volatile and semivolatile organic compounds (VOCs and SVOCs). With respect to a healthy indoor environment, only low emitting products, which do not influence the indoor air quality in a negative way, should be used in a building. Therefore, materials and products for indoor use need to be evaluated for their chemical emissions. Many emission studies have shown that the types of sources in occupational and residential indoor environments, the spectrum of emitting VOCs and SVOCs, the emission rate and the duration of emission cover a wide range. The demand for standardised test methods under laboratory conditions has resulted in several guidelines for the determination of emission rates by use of test chambers and cells. Furthermore, it is now recognised that both primary and secondary emissions may affect indoor air quality. As a consequence, modern product development should also consider secondary products, which seem to be of importance for long-term emissions. In order to characterise the release of VOCs and SVOCs from materials under realistic conditions, it is important to study the influence of processing, substrate and climatic parameters on emitting species and emission rates.

Keywords   Building products · Emission testing · Volatile organic compound · Semivolatile organic compound · Emission rate

1
Introduction

Building products, furnishings and consumer goods often demonstrate emissions of volatile chemical compounds during usage, and these can become a problem particularly under unfavourable indoor climate conditons. These emissions mostly involve volatile organic compounds (VOCs) or semivolatile organic compounds (SVOCs), the differentiation being based on the boiling points [1] or chromatographic properties [2]. The original causes for the occurrence of these emissions are solvents, residual monomers, plasticisers, flame retardants, process auxiliaries and preservatives (biocides), which are added to the previously mentioned products in order to achieve specific properties. Furthermore, there are also emissions which do not occur until during the utilisation phase. These include, for example, chemical reaction and decomposition products.

Indoor air measurements are often carried out on air polluting substances. However, the determination of the VOC and SVOC noxious emission situation is only the first stage in respect of achieving an effective reduction in emissions. Consequently the question must be raised regarding the specific emission sources and their contribution to the total pollution as a function of indoor concentration and climate influences such as temperature, humidity and air exchange. This is demonstrated in Fig. 1 for the example of a building product.

The evaluation of the emission potential of individual products and materials under realistic conditions and over defined timescales therefore requires the use of climate-controlled emission testing systems, so-called emission test chambers and cells, the size of which can vary between a few cubic centimetres and several cubic metres, depending on the application. The selection of the systems, the sampling preparation and the test performance all depend on the task to be performed.

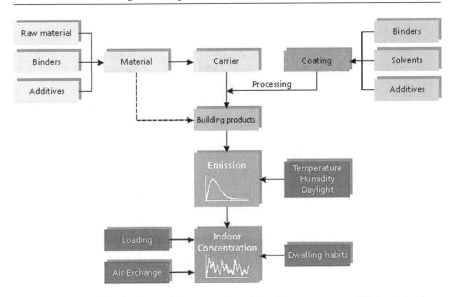

Fig. 1 Influence of applied materials, processing, climatic parameters and living behaviour on emissions and indoor concentrations

Nowadays emission measurements are performed for the following purposes [3]:

- Compiling of substance-specific emission data from various sources to back up field investigations into indoor air quality.
- Determination of the influence of environmentally relevant factors such as temperature, humidity and air exchange on the emission characteristics of the products.
- Processing of characteristic emission data to estimate product emissions.
- Processing of characteristic emission data to develop models that can be used to predict indoor concentrations.
- Ranking of various products and product types on the basis of the characteristic emission data.

On the basis of characteristic emission data there is an increasing tendency to award a quality label to particularly "low emission" products [4].

2
Equipment for Emission Testing

2.1
Emission Test Chamber

The first large-scale test chambers with a volume of around 40 m³ were developed in the mid-1970s in the course of the introduction of building authority

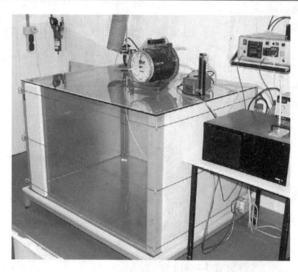

Fig. 2 View of a 1-m³ emission test chamber used in WKI (Photo: T. Salthammer)

regulations concerning formaldehyde emissions from wood particle boards. These were used to carry out close-to-practice formaldehyde measurements on particle board at a ratio of material surface area to chamber volume of 1 m² m⁻³, a temperature of 23 °C, a relative humidity of 45 % and an air exchange rate of 1 h⁻¹. Interest soon expanded to include VOCs and nowadays chamber investigations are probably the most important method used for the determination of the emission characteristics of materials. Moreover, a reduction in the chamber size had to be achieved for practical reasons. It is true that large chambers permit realistic investigations on large-area emittents with a complex composition, such as furnishing units, but these are impractical from both apparatus and time-consumption aspects. In principle a differentiation can be made between the so-called small-scale chambers with a volume in the range from a few litres to several cubic metres and the large-scale chambers ("walk-in type"). A self-manufactured (WKI) 1-m³ glass chamber [5] is shown in Fig. 2.

2.1.1
Investigation Principle

During an emission investigation the product/material to be investigated is tested with regard to temperature, relative humidity, air exchange rate, air velocity and product loading factor (ratio of surface of product to be investigated to the volume of the emission test chamber) under standardised testing conditions in an emission test chamber that can be sealed gas-tight against the outside atmosphere. The test procedure is suitable for emission investigations of both surfaces and of volume samples. This is a convention process where the boundary conditions are selected in such a way that they reflect those to be found in realistic indoor rooms. In interpreting the results of test chamber investigations it must, under certain circumstances, be taken into account as a limiting factor that not all realistic conditions to be found in an indoor room can be simulated.

2.1.2
Sink Effects

A sink effect is the fact that the components released partially adsorb within the test chamber, for example, at the chamber walls [6]. This can result in an incorrect reading of the concentration determined at the chamber outlet, which can lead to the wrong emission rate being computed. In principle every test chamber demonstrates a (low) sink effect. The degree of adsorption on chamber walls and in materials as well as the extent of the recovery can vary extremely for different VOCs [5, 7]. This depends in part on the volatility of the relevant substance. Put simply, it can be stated that absorption within the test chamber is favoured as the boiling point increases. A comparatively high polarity of a compound can also be favourable to the adsorption at the chamber walls [8]. Sollinger and Levsen [6] also showed that the sink effect is clearly increased by introducing an adsorbing sample surface (carpet) into the test chamber. The fact that the sample to be investigated in a chamber investigation also acts as a sink has been reported by several authors [5, 9, 10].

Whereas the sink effects caused by the chamber itself can be reduced to a minimum by using appropriate construction materials, it is not possible to influence the sink effects attributable to the actual sample. Many materials subjected to emission testing are also good sinks for the substances emitted by them. Especially with porous building materials with a large surface area or with foams it must be expected that the intrinsic material sink effect leads to a clear delay of the substance emission. An experiment [5] where the test substance 1,2,3-trichlorobenzene (1,2,3-TCB) was charged into the test chamber continuously over a period of 3 days (sorption phase) is shown in Fig. 3. The increase in the chamber concentration and the decay after the charging was completed (desorption phase) at an air exchange rate of 1 h^{-1} were measured by sampling with Tenax. The experiment was then repeated using a building panel in the chamber. This revealed that both the increase and the decay of the concentration of the test substance were considerably delayed. Whereas with the empty chamber the 1,2,3-TCB concentration dropped to almost zero 300 min after the charging was complete, this process lasted more than 3 days with the sink in the chamber. The sink effect of the sample material therefore means that the measured emission rate is smaller than the actual rate, as emitted substances are initially absorbed in the sink. At the same time the measured duration of the emission process is longer, as even after the completion of the actual emission substances are still released from the sink and lead to the chamber concentration dropping more slowly.

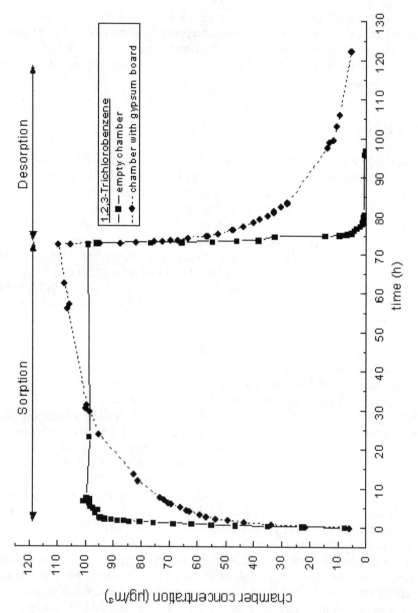

Fig. 3 Concentration versus time for 1,2,3-trichlorobenzene in the empty chamber and with gypsum board. Source Uhde [5]

This sink effect caused by the sample to be investigated plays a role in each chamber investigation and cannot be avoided. For this reason it is of fundamental importance to guarantee that the chamber has as good a recovery as possible, which is generally determined for selected VOCs in a concentration range that is typical for an emission investigation using one of the following processes: charging of test gases with known VOC content into the chamber, placing of permeation or diffusion tubes in the test chamber or regular dosing of the VOCs by forced injection within the test chamber [11]. The VOC concentration is measured at the purge gas outlet and compared with the theoretical target value. According to the European Standard EN 13419-1, the recovery for product-specific VOCs (toluene, n-dodecane) should be more than 80% [12].

In connection with the sink effect it should also be taken into consideration that a so-called memory effect can result for subsequent investigations, and this must be avoided by appropriate cleaning of the chamber. It is advantageous here if the test chamber can be thoroughly heated through at higher temperatures for cleaning purposes [8].

The combination of the air exchange rate and the product loading factor results in the area-specific flow rate, q, with the unit cubic metres per hour per square metre. It is recommended that building products are investigated at the surface-specific flow rates, which reflect the actual conditions within buildings. Where the q values are too low, this can lead to the actual emission rates being underestimated [13]. In the investigation of floor coverings it is proposed, for example, that a value of $1.25 \ m^3 \ h^{-1} \ m^{-2}$ is selected [4]; however it must be considered here that in a nonsteady state different concentrations can result, although the combinations of various air exchange rates and product loading factors produce the same value for the area-specific flow rate [14].

2.2
Emission Test Cell

It is often desired to have a device for a quality control of products with regard to emissions arising during the utilisation phase. To enable action to be taken during the production process as quickly as possible, fast emission investigations accompanying the production process would be a worthwhile objective. Owing to the high apparatus expenditure involved in test chamber investigations many manufacturers are forced to send their postproduction samples to testing institutes that possess the relevant analytical equipment. The result of the investigation is therefore only available after some delay, mostly after the product investigated has left the production facilities. It is therefore desirable to have a measuring system that can be used to carry out emission testing and quality assurance on location. The relevant principle of a transportable emission testing cell for mobile application was implemented in Scandinavia for the first time in 1991 with the so-called field and laboratory emission cell [15–18], see also Fig. 4. The field and laboratory emission cell opens up the opportunity

Fig. 4 Field and laboratory emission cell – top view (Photo: T. Salthammer)

of carrying out nondestructive emission testing on surfaces within the framework of field investigations. In this way it is possible to identify emissions from building products when they are already installed and also the sources for air-polluting substances.

2.3
Sampling and Analysis

The state of the art in emission investigations is nowadays discontinuous active sampling on appropriate adsorbents [19]. Using two different processes a great number of relevant VOCs can be determined. Sampling with Tenax allows enrichment of polar and nonpolar compounds with boiling points above 60 °C. The identification is carried out by thermal desorption using gas chromatography/mass spectrometry, the quantification using gas chromatography/mass spectrometry or gas chromatography/flame ionisation detection. The dinitrophenyl hydrazine (DNPH) process has become established for aldehydes and ketones. Here the substances are derivated to hydrazones and quantified using high-performance liquid chromotagraphy with UV detection. Further sampling and analysis techniques are used in individual cases to determine very volatile organic compounds, SVOCs or reactive organic compounds. The processes referred to are generally highly sensitive and precise, but they are time-intensive and require a great deal of apparatus. However a total parameter might be sufficient for quality control accompanying the production process. Continuous recording devices on the basis of flame ionisation, photoionisation or photoacoustics are suitable in principle. A uniform procedure was proposed at European level [4] for the determination of a discontinuously ascertained VOC total value (TVOC). Detailed treatment of VOC/SVOC sampling and analysis in indoor air and in test chambers can be found in Refs. [20, 21].

2.4
Calculation of Emission Rates

Air measurement in a chamber or cell initially produces the concentration $C(t)$ at time t of the measurement. $C(t)$ is, however, dependent on many variables, particularly on the air exchange rate and the loading factor, and possibly on the temperature, humidity, and air speed at the sample surface. To enable better comparability of the measured data the specific emission rate (SER) independent of air exchange and loading is to be preferred. The SER describes the product-specific emission behaviour for selected chemical compounds or for the TVOC, for example, as surface-specific emission rate with the unit micrograms per square metre per hour or as a unit-based SER (SER_U) with the unit micrograms per unit per hour. The SER can be used to compare various products with each other (ranking) and to designate particularly low emission products (labelling).

The time-dependent determination of the emission potential is carried out according to the following balance (Eq. 1), where $C(t)$ is the chamber concentration in micrograms per cubic metre, n is the air exchange per hour and L is the loading in square metres per cubic metre:

$$dC/dt = L SER(t) - nC(t). \tag{1}$$

For a decaying concentration–time function $SER(t)$ is obtained from (Eq. 1) by transition to differential quotients according to Eq. (2):

$$SER(t) = [(\Delta C/\Delta t) + nC(t)]/L, \tag{2}$$

with

$$\Delta C_i/\Delta t_i = [(C_i - C_{i-1})/(t_i - t_{i-1}) + (C_{i+1} - C_i)/(t_{i+1} - t_i)]/2. \tag{3}$$

Thus, if there are $n+1$ data points for concentration, $n-1$ emission rate values can be obtained by this method [21]. In the steady state ($dC/dt=0$) Eq. (2) progresses to Eq. (4):

$$SER = (nC)/L. \tag{4}$$

In ASTM 5116 [22] it is pointed out that the estimation of the SER from Eq. (4) may have a significant error if the emission rate is not constant or if the chamber has not reached steady state.

More sophisticated emission source models can be found in the literature [23]. Calculation of emission rates requires nonlinear curve fitting with a sufficient number of data points. Large errors in parameter estimates can result from rough chamber data and wrong models [22, 24].

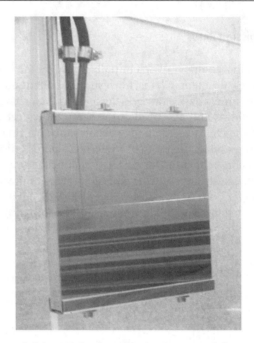

Fig. 5 View of the cooled (15 °C) fogging plate (15 cm×15 cm) for sampling semivolatile organic compounds. Source Uhde et al. [71]

2.5
Fogging Sampling

In emission studies the release of SVOCs from building products is also of interest. To assess the condensable amount of SVOCs, a fogging apparatus can be mounted in a chamber. The device described here was developed by Bauhof and Wensing [25] for SVOC measurement in automobile interiors. It is made of a cooled aluminium body (15 °C at 23 °C) with highly polished stainless steel collection plates mounted on each side (Fig. 5). After exposure to the chamber atmosphere the plates are dismounted and put face to face into a stainless steel extraction apparatus, where the condensed substances are eluted in an ultrasonic bath. The extracts are then analysed. The fogging value (in micrograms) determined over a period of 14 days is a characteristic value for the amount of SVOC that can be expected to condense on cooler surfaces in an indoor environment. The fogging method is based on a convention. The comparison of results requires identical experimental performance. In contrast to air concentrations, which are a measure for an inhalative exposition, the fogging value represents an oral or dermal exposition. It allows for an estimation of the amount of target compounds that may condense on cool surfaces or porous material (like house dust or textiles) and therefore may enter the body through skin contact or ingestion.

2.6
Sample Preparation for Emission Testing

In selecting parameters for emission testing and interpreting the results it must be taken into consideration that this is a convention process and that not all real conditions in indoor rooms can be simulated. The results of such investigations can be increasingly used in the future to develop low-emission products and, by selecting such products in the form of source-checking, make a considerable contribution to achieving good indoor air quality.

The so-called sample preparation is of particular importance with regard to emission test results. This involves, for example, the question whether liquid samples (paints, adhesives, etc.) are to be applied on absorbent or inert substrates. Investigations of paints have found that clearly different emissions occur depending on the substrate [26]. A sample preparation as close to the realistic situation as possible is always appropriate where the results are to be used to derive health-related statements. In investigating floor coverings it can be useful to investigate only the product surface that is exposed to the indoor room and to seal off the backing and any side edges during the emission test with materials with as low as possible emission. To achieve as realistic an examination as possible, emission investigations of complete sandwich systems, for example, lime flooring, levelling compound, primer, adhesive and top layer, are of particular interest. Such test scenarios can provide valuable information on compatibility and possible chemical reactions between the individual building products. If only the individual products are subjected to separate emission testing procedures, then follow-up reactions, such as the hydrolytic decomposition of an adhesive, cannot be found owing to the alkalinity of the lime flooring [27].

2.7
Standardisation of Test Methods

The result of an emission investigation is influenced by a number of different factors (Table 1). In order to ensure comparability of the relevant investigations the three stages of the process, namely, sample preparation, emission measurement and chemical analysis, must be standardised. This is also shown by earlier results from relevant comparative investigations [28, 29].

In connection with building products directives, general demands have been elaborated for VOC emission investigations of building products and include

– Investigation of building products using emission test chambers.
– Investigation of building products using emission test cells.
– Sampling of building products and sample preparation.
– VOC analysis.

In this standardisation work, which has already been reported in Ref. [30], it was possible to fall back on important previous works [3, 28, 29, 31, 32]. EN, ISO

Table 1 Influencing factors in an emission investigation. *VOC* represents volatile organic compound and *SVOC* represents semivolatile organic compound

Product history	Age of product
	Transport to the test laboratory and storage time
	Sample preparation
	Climatic conditioning prior to testing
Chamber measurement	Supply air quality
	Background concentration
	Sink effects/recovery
	Air exchange ratio
	Air tightness of the environmental test chamber
	Internal air mixing
	Air velocity
	Accuracy of temperature, relative humidity,
	air exchange ratio, product loading factor
Sampling of test chamber air and chemical analysis	Choice of sorbent tube
	Recovery of specific VOC and SVOC
	Detection limit
	Precision
	Accuracy

Table 2 Standard methods for emission testing

Standard	Description	Reference
EN 13419-1	Emission test chamber	[12]
EN 13419-2	Emission test cell	[33]
EN 13419-3	Preparation of test specimens	[34]
ISO 16000-6	Determination of VOCs in chamber air	[35]
ASTM D 5116	Small-scale chamber	[22]
ASTM D 6670	Full-scale chamber	[36]

and ASTM standards on emission testing are summarised in Table 2. An overview of the quality demands of EN 13419-1 [12] and EN 13419-2 [33] for important parameters in VOC emission investigations using emission test chambers and cells is provided in Table 3. Abbreviations and definitions according to EN 13419 are summarized in Appendixes 1 and 2. ASTM standards on indoor air quality are compiled in Ref. [37].

Table 3 Quality requirements for emission test chamber and emission cell according to EN 13419-1 [12] and EN 13419-2 [33]. *TVOC* represents total volatile organic compound

Parameter	Emission test chamber	Emission cell
Temperature	23±1 °C	23±1 °C
Relative humidity	50±5%	50±5%
Air velocity	0.1–0.3 m s^{-1}	0.003–0.3 m s^{-1}
Air tightness	Leak rate <1% of supply air	Difference between supply air and exhausted air <3%
Background value	<10 μg m^{-3} TVOC, <2 μg m^{-3} single compound	<10 μg m^{-3} TVOC, <2 μg m^{-3} single compound
Recovery	≥80% toluene, *n*-dodecane	≥90% toluene, *n*-dodecane
Air mixing	±10% of the theoretical model of total mixing	
Air exchange rate	Accuracy ±3%	Accuracy ±3%

3
Emission Studies

Most materials and products for indoor use contain organic compounds, which can be released in an indoor environment. The majority of the VOCs described in the last few years are typical industrial chemicals that are emitted from building materials and furnishings, that are not reactive under indoor room conditions and that can be essentially assigned to the following groups:

- Aliphatic hydrocarbons.
- Aromatic hydrocarbons.
- Alcohols.
- Ketones.
- Esters.
- Glycols/glycol ether.

To date most indoor air investigations and the quality demands for indoor products have been limited to these compounds, as the potential emission sources have been documented in detail and sensitive analytic methods are available as a matter of routine. In contrast only a few VOCs of relevance indoors have been characterised until now as "reactive" or as "reaction products". Some groups are as follows:

- Unsaturated hydrocarbons.
- Terpenes.
- Aldehydes.
- Organic acids.
- Acrylate monomers.
- Diisocyanate monomers.

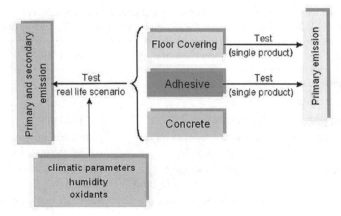

Fig. 6 Influence of test conditions on primary and secondary emission (substrate–adhesive–flooring material)

However, it is particularly such components that can have an influence on human well-being even in low concentrations. In a recent publication Wolkoff et al. [38] contended that especially reactive organic compounds and their reaction products can cause irritation in persons and therefore make a significant contribution to the sick building syndrome. Many building materials, furnishings and household products have nowadays been documented as sources of reactive and nonreactive VOCs. Furthermore, the results of research work have shown that "indoor chemistry" is of great importance in the evaluation of indoor air quality. Various products contain reactive components which decompose under the influence of oxidants, heat, moisture or light in the material [39]. In many cases volatile compounds result in this way with a sensory effect [40] and are released into the room air as secondary emissions. Case studies have shown that particularly certain combinations in floor composites (e.g. floor topping–adhesive–top cover) are to be regarded as a potential source of secondary emissions [27]. In addition new VOCs can be formed even during the manufacturing process. As a consequence, emission testing of single products under standard conditions will only yield the primary products. A real-life scenario is required to see both primary and secondary emissions (Fig. 6). A list of relevant products and secondary emissions is given in Ref. [41].

Tucker [42] has summarised the major factors that are now thought to influence emission of vapour-phase organic compounds from surface materials:

- Total amount and volatility of constituents in the material.
- Distribution of these constituents between the surface and the interior of the material.
- Time (i.e. age of the material).
- Surface area of the material per volume of the space it is in ("loading").

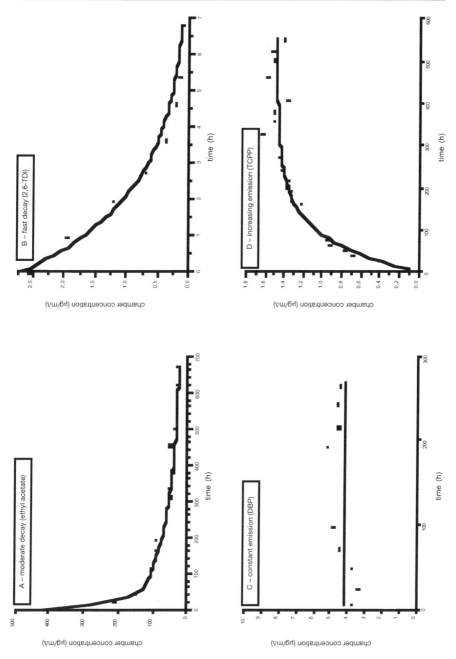

Fig. 7 Emission profiles: **a** moderate decay, emission of ethyl acetate from UV-cured lacquer; **b** fast decay, emission of 2,6-toluene diisocyanate from construction foam; **c** constant emission, emission of di-n-butyl phthalate from wallcovering; **d** increasing emission, emission of tris(chloropropyl)phosphate from a television set. Test conditions 1-m^3 chamber, 23 °C, 50% relative humidity, n=1 h^{-1}

- Environmental factors such as temperature, air exchange rate and relative humidity.
- Chemical reactions in the source (e.g. conversion in varnishes and some adhesives).

Therefore, VOC and SVOC emission profiles may differ depending on the factors mentioned as is demonstrated in Fig. 7 for moderate decay, fast decay, constant emission and increasing emission. All the studies were performed at 23 °C and 50% relative humidity. Even for the same type of products, the emission may be different over several orders of magnitude as is demonstrated in Fig. 8 for UV-cured furniture coatings. Here, chamber concentrations (Σ VOC) were measured under identical conditions after 24 h and 672 h, respectively (see Ref. [43] for details).

Tucker [42] also compiled emission factors (SER_A) for various compounds from a variety of sources as reported in the literature. Note that SER_A values vary by a factor of many thousands from one material to another. A summary of potential sources for VOCs and SVOCs in an indoor environment is provided in Table 4.

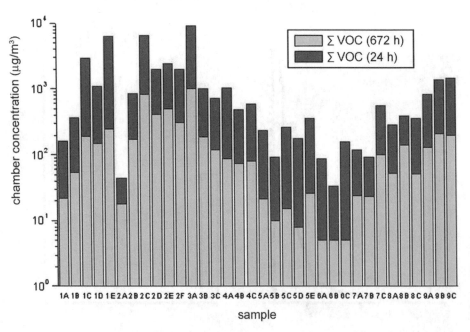

Fig. 8 Σ VOC values (μg m^{-3}) for emission from UV-cured lacquer after 24 h and 672 h. Test conditions 23 °C, 50% relative humidity, $q=1$ m^3 m^{-2} h^{-1}

Table 4 Compilation of some VOC/SVOC emission studies on building products and indoor materials. *FLEC* represents field and laboratory emission cell

Products	Description	Authors	Ref.
Latex paint	Influence of substrate on VOC emission	Chang et al. (1997)	[26]
Latex paint	Study of VOC emission (chamber)	Sparks et al. (1999)	[44]
Latex paint	Study of VOC emission (model house)	Sparks et al. (1999)	[45]
Latex paint	Influence of ozone on VOC emission	Reiss et al. (1995)	[46]
Water-based paint	Influence of film thickness on VOC and SVOC emission	Clausen (1993)	[47]
Water-based paint	VOC and TVOC emission study (chamber)	Wensing (1996)	[48]
UV-curing lacquer	VOC emission in dependency of climatic parameters (chamber, FLEC)	Salthammer and coworkers (1999, 2002)	[49, 50]
Wood	VOC emission (FLEC)	Englund (1999)	[51]
Biocides	Long-term emission of dichlofluanide, permethrine und tebuconazole (chamber)	Jann and Wilke (1999)	[52]
PVC flooring	VOC emission (FLEC)	Lundgren et al. (1999)	[53]
Particle board	VOC and formaldehyde emission (chamber)	Brown (1999)	[54]
Polyurethane products	Emission of 2,4-toluene diisocyanate and 2,6-toluene diisocyanate I (chamber)	Kelly et al. (1999)	[55]
Polyurethane adhesives	Emission of 4,4′-diphenylmethane diisocyanate and hexamethylene diisocyanate, influence of temperature, monomer content, curing mechanism	Wirts and Salthammer (2002)	[56]
Cork	VOC emission (chamber)	Horn et al. (1998) Salthammer and Fuhrmann (2000)	[57] [58]
Flooring materials (miscellaneous)	Compilation of VOC emission studies	Saarela (1999)	[59]
Building products (miscellaneous)	Compilation of VOC emission studies	Tucker (2001)	[42]
Building products (miscellaneous)	Influence of temperature on VOC emission	Van der Wal et al. (1997)	[60]
Building products (miscellaneous)	VOC emission study (chamber and FLEC)	Zellweger et al. (1997)	[61]
Building products (miscellaneous)	Dependence of degradation on building products (chamber)	Moriske et al. (1998)	[62]
Building products (miscellaneous)	VOC emission in dependence of climatic parameters	Wolkoff (1998)	[39]

Table 4 (continued)

Products	Description	Authors	Ref.
Linoleum	VOC emission (FLEC)	Jensen et al. (1995)	[63]
Linoleum	Sensory testing	Jensen et al. (1995)	[64]
Furniture surface coating	VOC emission (chamber)	Salthammer (1999)	[65]
Wooden furniture	Emission of terpenes (chamber and FLEC)	Salthammer and Fuhrmann (1996)	[66]
Adhesive	Emission of toluene (chamber and model room)	Nagda et al. (1995)	[67]
Copy machine, laser printer and PCs	Emission of VOC (TVOC), ozone and particles (chamber)	Black and Worthan (1999)	[68]
Television and video sets	VOC and SVOC emission, new and aged products (chamber)	Wensing (1999)	[69]
Textile floor covering, PVC flooring, water-based paint	Round-robin test (chamber and FLEC)	De Bortoli et al. (1999)	[70]
Wall coverings	SVOC emission (chamber)	Uhde et al. (2001)	[71]
Textile floor coverings	VOC emission (chamber)	Solllinger and Levsen (1993, 1994)	[6, 72]
Building products (miscellaneous)	Sensory emission, impact of climatic parameters	Knudsen and coworkers (1997, 1999) Fang et al. (1999)	[40, 73] [74]
Building products (miscellaneous)	Sorptive interactions between VOC and materials	Meininghaus and coworkers (1999) Won et al. (2001)	[7, 75] [76]

3.1
Solvents

The wide variety of common solvents and film formers (aromatic hydro-carbons, aliphatic hydrocarbons, ketones, esters, glycols, alcohols, Table 5) still represents a very important group of emitting compounds, as freshly manu-factured products may cause high SER_A values. On the other hand, they do not have any chemical reactions under normal conditions, unless so-called reactive solvents (e.g. styrene) are involved. An important exception is 2,4,7,9-tetra-methyl-5-decyne-4,7-diol (T4MDD), which is frequently used as a film former and foam inhibitor in water-based coating systems. T4MDD degrades to 4-methyl-2-pentanone (MIBK) and 3,5-dimethyl-1-hexyne-3-ol. As MIBK is a common solvent the degradation of T4MDD can be mistaken for the use of

Table 5 Examples of compounds detected in indoor air and from building products in test chambers (see also Refs. [2, 35])

Compound	CAS number	Remarks
Aromatic compounds		
Benzene	71-43-2	Impurity
Toluene	108-88-3	Solvent
Ethylbenzene	100-41-4	Solvent
m-Xylene	108-38-3	Solvent
p-Xylene	106-42-2	
o-Xylene	95-47-6	Solvent
Isopropylbenzene	98-82-8	Solvent
n-Propylbenzene	103-65-1	Solvent
1,2,4-Trimethylbenzene	95-63-6	Solvent
1,3,5-Trimethylbenzene	108-67-8	Solvent
1,2,3-Trimethylbenzene	526-73-8	Solvent
1,2,4,5-Tetramethylbenzene	95-93-2	Solvent
1-Methyl-2-propylbenzene	1074-17-5	Solvent
1-Methyl-3-probylbenzene	1074-43-7	Solvent
n-Butylbenzene	104-51-8	Solvent
1,3-Diisopropylbenzene	99-62-7	
1,4-Diisopropylbenzene	100-18-5	
2-Phenyldecane	4537-13-7	
5-pPhenyldecane	4537-11-5	
5-Phenylundecane	4537-15-9	
o-Methylstyrene	611-15-4	
m-Methylstyrene/*p*-methylstyrene	100-80-1/622-97-9	
α-Methylstyrene	98-83-9	Monomer
2-Ethyltoluene	611-14-3	
Styrene	100-42-5	Solvent, monomer
Naphthalene	91-20-3	
4-Phenylcyclohexene (styrene–butadiene rubber)	31017-40-0	Diels–Alder product
Aliphatic hydrocarbons		
n-Hexane	110-54-3	Solvent
2-Methylhexane	591-76-4	
3-Methylhexane	589-34-4	
n-Heptane	142-82-5	Solvent
n-Octane	111-65-9	Solvent
n-Nonane	111-84-2	Solvent
2-Methyloctane	3221-61-2	
3-Methyloctane	2216-33-3	
2-Methylnonane	871-83-0	
3,5-Dimethyloctane	15869-93-9	
n-Decane	124-18-5	Solvent
2,4,6-Trimethyloctane	62016-37-9	
4-Methyldecane	2847-72-5	

Table 5 (continued)

Compound	CAS number	Remarks
Aliphatic hydrocarbons		
n-Undecane	1120-21-4	Solvent
2,2,4,6,6-Pentamethylheptane	13475-82-6	
n-Dodecane	112-40-3	
n-Tridecane	629-50-5	
n-Tetradecane	629-59-4	
n-Pentadecane	629-62-9	
n-Hexadecane	544-76-3	
2-Methylpentane	107-83-5	
3-Methylpentane	96-14-0	
1-Octene	111-66-0	
1-Decene	872-05-9	
Cyclic hydrocarbons		
Methylcyclopentane	96-37-7	
Cyclohexane	110-82-7	Solvent
1,4-Dimethylcyclohexane	70688-47-0	
4-Vinylcyclohexene	100-40-3	Reaction product (styrene–butadiene rubber)
Methylcyclohexane	108-87-2	
Terpenes		
Δ^3-Carene	13466-78-9	Wood
α-Pinene	80-56-8	Wood, eco-lacquer
Camphene	79-92-5	
β-Pinene	18172-67-3	Wood, eco-lacquer
Longifolene	475-20-7	Wood
α-Cedrene	469-61-4	Wood
Caryophyllene	87-44-5	Wood
Limonene	138-86-3	Wood, eco-lacquer
Alcohols		
1-Propanol	71-23-8	Solvent
2-Propanol	67-63-0	Solvent
2-Methyl-2-propanol	75-65-0	
2-Methyl-1-propanol	78-83-1	
1-Butanol	71-36-3	Solvent
1-Pentanol	71-41-0	
1-Hexanol	111-27-3	
Cyclohexanol	108-93-0	Solvent
1-Octanol	111-87-5	
2-Ethyl-1-hexanol	104-76-7	Degradation product [di-2-(ethylhexyl)phthalate]
Phenylmethanol	100-51-6	
Phenol	108-95-2	
Butylated hydroxytoluene	128-37-0	Stabiliser

Table 5 (continued)

Compound	CAS number	Remarks
Glycols/glycol ethers		
1,2-Propylene glycol	57-55-6	
Dimethoxymethane	109-87-5	
Dimethoxyethane	110-71-4	
Diethylene glycol-*n*-mono-butyl ether	112-34-5	
2-Methoxyethanol	109-86-4	
2-Ethoxyethanol	110-80-5	Solvent
2-Butoxyethanol	111-76-2	Solvent
1-Methoxy-2-propanol	107-98-2	
2-Butoxyethoxyethanol	112-34-5	
2-Phenoxyethanol	122-99-6	Solvent
Aldehydes		
Acetaldehyde	75-07-0	
Propanal	123-38-6	
Butanal	123-72-8	Degradation product (fatty acids)
Pentanal	110-62-3	Degradation product (fatty acids)
Hexanal	66-25-1	Degradation product (fatty acids)
Heptanal	111-71-7	Degradation product (fatty acids)
2-Ethylhexanal	123-05-7	
Decanal	112-31-2	Degradation product (fatty acids)
2-Pentenal	1576-87-0	Degradation product (fatty acids)
2-Heptenal (trans)	18829-55-5	Degradation product (fatty acids)
2-Nonenal (trans)	18829-56-6	Degradation product (fatty acids)
2-Undecenal (trans)	53448-07-0	Degradation product (fatty acids)
Octanal	124-13-0	Degradation product (fatty acids)
2-Butenal	123-73-9	Degradation product (fatty acids)
2-Furancarboxaldehyde (furfural)	98-01-1	Degradation product (cork)
Nonanal	124-19-6	Degradation product (fatty acids)
Benzaldehyde	100-52-7	Degradation product (UV, Tenax)
2,4,6-Trimethylbenzaldehyde	487-68-3	Degradation product (UV, Tenax)
Ketones		
Acetone	67-64-1	Solvent, degradation product (UV)
2-Butanone (MEK)	78-93-3	Solvent
3-Methyl-2-butanone	563-80-4	
4-Methyl-2-pentanone (MIBK)	108-10-1	Solvent, degradation product
Cyclopentanone	120-92-3	
2-Methylcyclohexanone	583-60-8	
2-Methylcyclopentanone	1120-72-5	
Cyclohexanone	108-94-1	Solvent, degradation product (UV)
Benzophenone	119-61-9	Photoinitiator
Acetophenone	98-86-2	Degradation product

Table 5 (continued)

Compound	CAS number	Remarks
Halogenic compounds		
Dichloromethane	75-09-2	Solvent
Carbon tetrachloride	56-23-5	Solvent
1,2-Dichloroethane	107-06-2	Solvent
Trichloroethene	79-01-6	Solvent
Tetrachloroethene	127-18-4	Solvent
1,1,1-Trichloroethane	71-55-6	Solvent
1,4-Dichlorobenzene	106-46-7	Moth crystal cake
Acids		
Acetic acid	64-19-7	Wood, wood-based materials
Propanoic acid	79-09-4	
Isobutyric acid	79-31-2	
Butyric acid	107-92-6	Degradation product (fatty acid)
2,2-Dimethylpropanoic acid	75-98-9	
Pentanoic acid	109-52-4	Degradation product (fatty acid)
Heptanoic acid	142-62-1	Degradation product (fatty acid)
Octanoic acid	124-07-2	Degradation product (fatty acid)
Hexadecanoic acid	57-10-3	
Esters		
Vinyl acetate	108-05-4	Monomer
Isopropyl acetate	108-21-4	Solvent
Isobutyl acetate	110-19-0	Solvent
Ethyl acetate	141-78-6	Solvent
Propyl acetate	109-60-4	Solvent
Butyl acetate	123-86-4	Solvent
2-Methoxypropyl acetate	108-65-6	Solvent
2-Methoxyethyl acetate	110-49-6	Solvent
2-Ethoxyethyl acetate	111-15-9	Solvent
2-Ethylhexyl acetate	103-09-3	Solvent
Texanol	25265-77-4	Solvent, plasticiser
TXIB	6846-50-0	Solvent, plasticiser
Tripropylene glycol diacrylate	42978-66-5	Monomer
Butyl acrylate	141-32-2	Monomer
2-Ethylhexyl acrylate	103-11-7	Monomer
Ethyl acrylate	140-88-5	Monomer
Phthalates		
Dimethyl phthalate	131-11-3	Plasticiser
Dibutyl phthalate	84-74-2	Plasticiser
Diisobutyl phthalate	84-69-5	Plasticiser
Di-2-ethylhexyl phthalate	117-81-7	Plasticiser
Di-phenyl phthalate	84-62-8	Plasticiser
Diisononyl phthalate	68515-98-0	Plasticiser

Table 5 (continued)

Compound	CAS number	Remarks
Phosphororganic compounds		
Triehylphosphate	78-40-0	Flame retardant
Tris(2-chloroethyl)phosphate	115-96-8	Flame retardant
Tris(chloropropyl)phosphate	13674-84-5	Flame retardant
Tris(dichloropropyl)phosphate	13674-87-8	Flame retardant
Tributylphosphate	126-73-8	Flame retardant
Miscellaneous		
1,4-Dioxane	123-91-1	Solvent
n-Methyl-2-pyrrolidone	872-50-4	Solvent
Methyl ethyl ketoxim (MEKO)	96-29-7	Antiskinning agent
Caprolactam	105-60-2	
Indene	95-13-6	
Tetrahydrofuran	109-99-9	Solvent
2-Pentylfuran	3777-69-3	
Hexamethylene diisocyanate	822-06-0	Monomer
4,4'-Diphenylmethane diisocyanate	101-68-8	Monomer
2,4-Toluene diisocyanate	584-84-9	Monomer
2,6-Toluene diisocyanate	91-08-7	Monomer
Tripropylene glycol diacrylate	42978-66-5	Monomer
Butyl acrylate	141-32-2	Monomer

MIBK in the analysis [41]. In test chamber investigations hemiacetals and acetals such as 1-hydroxy-1-ethoxycyclohexane and 1,1-diethoxycyclohexane can sometimes be demonstrated qualitatively in the chamber air. These are typical reaction products of cyclohexanone with ethanol.[1]

3.2
Aldehydes

Saturated and unsaturated aldehydes with chain lengths C5–C11 are some of the most problematic and undesirable substances in indoor rooms. Aliphatic aldehydes are very odour-intensive, and the odour is generally described as unpleasantly "rancid" or "greasy". With sensitive persons or in high concentrations the perception of aliphatic aldehydes can cause nausea. Odour thresholds are, for example, 57 and 13 µg m^{-3} [77] for hexanal and nonanal, and the odour thresholds of the unsaturated aliphatic aldehydes are even a magnitude smaller. Emission sources in indoor rooms are essentially unsaturated fatty acids such as oleic acid, linoleic acid and linolenic acid as components of

---

[1] A known artefact in gas-chromatograph analysis is the formation of 1,1-dimethoxy-cyclohexane from cyclohexanone and methanol in standard solutions.

linoleum [63,64], coating systems containing alkyd resin [78] as well as paints, oils and adhesives on a natural basis. The aldehydes generated during the oxidation processes of the unsaturated fatty acids can continue to react to the corresponding low-chain acids. Typical degradation products of oleic acid are saturated aldehydes from heptanal (C7) to decanal (C10), whereas linoleic acid is mainly degraded to hexanal (C6). Unsaturated aldehydes such as 2,4-hepta-dienal result from oxidising the double-unsaturated linolenic acid. The degra-dation of fatty acids was investigated particularly with regard to food chemistry aspects as aldehydes are strong aroma components [79]. This property is also used in so-called air-freshening sprays, which also contain saturated aldehydes [80]. The surface of a piece of furniture can under certain circumstances demonstrate high emission rates for aldehydes for months and even years. The reason for this lies in the constant slow degradation of unsaturated fatty acids under living conditions. It was observed with various furniture surfaces that odour problems can occur only after several years of use. In these cases a cracking of the paint film allows oxidative processes after the penetration of atmospheric oxygen.

Cork products can also emit considerable amounts of the aldehyde furfural (2-furancarboxaldehyde) with surface-specific emission rates $SER_A >$ 1,000 $\mu g \, m^{-2} \, h^{-1}$ [57], whereby the release of the furfural generally corresponds with that of acetic acid. The cause for the formation of furfural is the thermal decomposition of the hemicelluloses contained in the cork at temperatures above 150 °C, while acetic acid is formed by the separation of acetyl groups. Further by-products identified in the thermal treatment of natural cork were formic acid and hydroxymethylfurfural [58].

3.3
4-Phenycyclohexene and Other Reaction Products

During the manufacturing process of styrene–butadiene rubber the poly-merisation is stopped at a conversion rate of less than 90%. The residual monomers styrene and butadiene are removed by distillation. The odour-in-tensive compounds 4-phenylcyclohexene (from styrene and cis-butadiene) and 4-vinyl cyclohexene (from cis-butadiene and trans-butadiene) can be formed from the remaining monomers under the conditions of a thermally permitted $\pi 2_s + \pi 4_s$ Diels–Alder cyclic addition. During indoor air measurements, carried out in six different office rooms in each case 3 days after new carpets had been laid, concentrations of 4-phenylcyclohexene of 29–45 $\mu g \, m^{-3}$ could be ascer-tained. It is suspected that the emission of the trimer of 2-methyl-1-propene from a glued carpet is also caused by chemical reaction.

Alkylated dithiocarbamates with chain lengths C2–C6 are often added as polymerization auxiliaries for styrene–butadiene rubber. Under acidic conditions, these intermediately form the free thioacid, which degrades in turn in a reaction analogous to Hofmann degradation to amine and carbon disulphide.

3.4
Terpenes

Terpenes can be emitted in large amounts both by plants and by building products and items of furnishing [66]. The terpenes involve mostly odour-intensive natural materials that are made up of units of 2-methylbutane and 2-methyl-1,3-butadiene (isoprene) [81]. Of relevance for the indoor area are the monoterpenes (C_{10}) such as α-pinene, β-pinene, limonene and Δ^3-carene as well as the sesquiterpenes (C_{15}) such as β-caryophyllene and longifolene. Especially resin-rich softwoods such as pine and spruce are strong emission sources for monoterpenes. However wood, wood products and coatings are by far not the only emission sources for terpenes in indoor rooms. The substances are also to be found in cleansing agents, cosmetics, air fresheners, paints, lacquers and oils. Jann et al. [82] examined the emission behaviour of limonene, α-pinene, β-pinene and carvone from furniture surfaces coated with natural paint, whereby in part very high emission rates were established. A very interesting compound is carvone, which, on the one hand, is a component in natural terpene balsam and, on the other hand, it is also a possible oxidation product of limonene. Δ^3-Carene is generally not an emission product of coatings, as it is removed from the raw products owing to its known effect as a contact allergen.

3.5
Photoproducts

Photoinitiators start polymerisation processes and are therefore indispensable components in radiation-curing systems. Commonly used photoinitiators, which undergo various fragmentation processes such as the Norrish I reaction, the Norrish II reaction (intramolecular proton transfer) and intermolecular proton transfer with exciplex formation, are benzophenone, hydroxyacetophenone, benzil ketals, phosphine oxides, benzoin ethers, thiocyanates and others [83]. It is a problem of UV technology that uncontrolled fragmentation reactions and remaining residues can cause yellowing and the release of odour-generating materials. On the other hand UV coatings are extremely low emission and functionally practical systems when the processing conditions are perfect.

A typical odour-intensive breakdown product generated by α-cleavage (Norrish I reaction) is benzaldehyde.[2] Owing to recombination and reduction processes further secondary products such as benzil and acetone [50] result from the fragmentation. In principle, odourless photoinitiators can therefore certainly cause significant odour problems via their degradation products.

---

[2] Certain photofragments such as benzaldehyde are simultaneously typical degradation products of Tenax TA and can arise as artefacts or blind readings [84].

3.6
Plasticisers

In order to obtain the desired material properties, PVC products are treated with additives. With regard to the quantity, the phthalic esters, which are used as plasticisers, represent the most significant portion. In an indoor environment PVC occurs mostly in household products, floor coverings, wall coverings and electronic devices [69]. Since wall and floor coverings sometimes represent a major part of the surface area in a room [85], they might be a source for phthalic acid esters. The soft PVC used preferentially in wallpapers contains plasticiser portions of about 30%. These are mainly di-n-butyl phthalate (DBP), di-2-ethylhexyl phthalate (DEHP) and diisononyl phthalate (DINP).

In a recent study the emissions of several technically relevant phthalates from PVC-coated wallcoverings were measured in emission test chambers under standard room conditions. During a 14-day test period both the chamber air concentrations and the condensation on a cooled plate (fogging) were determined. Chamber concentrations of DBP reached a maximum of 5 µg m^{-3} (Fig. 7c). Correspondingly, the plasticisers diphenyl phthalate (DPP) and DEHP, which have higher boiling points, reached maximum concentrations of about 1–2 µg m^{-3}. After 14 days of exposure, up to 60.4 µg DEHP and 17.7 µg DPP

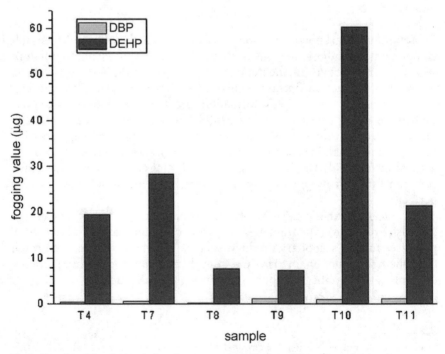

Fig. 9 Results of fogging tests in micrograms (sampled on 450 cm^2 over 14 days)

could be quantified on the cooled plates of the fogging apparatus. The amounts of DBP and DIBP were significantly lower (Fig. 9).

In 1996, investigations on PVC products were performed in test chambers under defined conditions [86]. At room temperature the DEHP concentration was below 0.1 µg m^{-3}. At higher temperatures the chamber values increased as expected. The maximum value was 5.2 µg m^{-3} (60 °C). Wilke and Jann [87] described test chamber investigations on wallpapers containing DEHP. In tests over 28 days in chambers of 1 m^3 (23 °C, 45% relative humidity, air exchange rate 1 h^{-1}, loading 1 m^2 m^{-3}) DEHP was not detectable with a detection limit of 0.35 µg m^{-3}. Only when the temperature was increased to 40 °C were DEHP values of up to approximately 2 µg m^{-3} measured.

A known and especially odour-intensive decomposition product from PVC-loaded materials is 2-ethyl-1-hexanol. The starting substance is DEHP. Under alkaline conditions and in the presence of water DEHP hydrolyses to the volatile 2-ethyl-1-hexanol and the anion of o-phthalic acid. The alkaline hydrolysis of DEHP occurs predominantly in floorings in conjunction with fresh floor topping. It is often the case that this still has a too high moisture content when the floor covering is laid [27]. This means that not only DEHP but also the flooring adhesive is degraded. In addition to mechanical problems (bubbling, peeling of the flooring) this leads to long-term emission sources of 2-ethyl-1-hexanol and further organic compounds which then mostly results in renovation being required. The investigation of a flooring adhesive on an inert base can therefore never supply information on the emission behaviour under real conditions. Fang et al. [74] have carried out release experiments of 2-ethyl-1-hexanol as a function of temperature and humidity for a carpet and a PVC flooring in a CLIMPAQ test chamber. As expected the surface-specific emission rate SER$_A$ clearly rose as the temperature and moisture increased. For the carpet the highest emission rate of 16.6 µg m^{-2} h^{-1} was measured at 28 °C and 70% relative humidity. DEHP is also photochemically degraded with the formation of 2-ethyl-1-hexanol and 2-ethylhexene [88].

3.7
Flame Retardants

Flame retardant is the collective term for those inorganic and/or organic compounds which give flameproof properties, particularly to wood and wood-working materials, plastics and textiles. They achieve this by reducing the inflammability of the material to be protected, preventing ignition and making burning difficult [89]. In regulating and implementing fire protection preventive measures, it is the safety aspect in the case of a fire that is taken into consideration in most cases. In practice, this means that the outbreak of an open fire with quickly spreading flames is what should be prevented as far as possible. Here is often not taken into consideration that during the burning process toxic products can be produced owing to thermal decomposition of flame retardants. These can contribute to contamination of the atmosphere or form persistent sediments. A further problem, widely ignored up to now, is the release of flame retardants and their decomposition products in an indoor environment under normal living conditions [90]. Appropriately treated construction materials, mattresses and textiles are notable in this. Electrical equipment is also equipped with flame retardants. The release of tris(chloro-propyl)phosphate from a television set is shown in Fig. 7d.

Many common flame retardants and their decomposition products are classified as dangerous substances or must be regarded as a potential health hazard and can cause impairment to health when inhaled, orally ingested or on contact with the skin. In the case of constructional measures or restoration work, the workers and the residents can be exposed to flame retardants, especially during use of construction foam.

In general, organic flame retardants can be roughly divided into three classes. The following overview shows a selection of the relevant compounds [91]:

- Halogenated organic compounds,
  - Chlorinated paraffins.
  - Polybrominated biphenyls.
  - Polybrominated biphenyl ethers.
- Nonhalogenated organophosphorus compounds.
- Halogenated organophosphorus compounds.

Exposure of residents to flame retardants mainly results from accumulation in house dust and indoor air. However, risk assessment for an indoor environment is difficult owing to the lack of emission studies, measured indoor air concentrations and toxicological data [92].

3.8
Monomers

In many polymerisation processes the viscosity, drying and other properties are controlled by monomers. These are introduced directly into the film as polymerisable thinners in the reaction with the reactive resin, whereby polyester acrylates or polyether acrylates can be formed depending on the components used. UV systems mainly use acrylate monomers, which are split up into monofunctional and polyfunctional systems in accordance with the number of $-CH=CH_2-$ groups in the end position. In today's application technology the highly volatile and odour-intensive monofunctional acrylates such as butyl acrylate and 2-ethylhexyl acrylate continue to enjoy less popularity in contrast to the multifunctional acrylates such as 1,6-hexandiol diacrylate, tripropylene glycol diacrylate (TPGDA) or trimethylol propane triacrylate. Free acrylate monomers can indeed have a negative influence on human well-being in indoor rooms as they cause irritations to the eyes and mucous membranes. Direct contact with acrylate monomers remaining on surfaces leads to skin irritation. A test of a freshly produced furniture surface yielded an initial

TPGDA concentration in the chamber air of 86 μg m^{-3}, which dropped to 2 μg m^{-3} within 500 hours, and after 600 hours the TPGDA could no longer be detected [93].

The nonacrylic monomer styrene is used as a reactive solvent in the polymerisation of coating systems on the basis of unsaturated polyesters. Styrene has a penetrating and unpleasant odour and a strong irritant effect. In spite of this substitute, products such as 2-phenyl-1-propene (α-methylstyrene), n-vinylpyrrolidone and vinyltoluene have not become established owing to their clearly inferior application properties.

The diisocyanates are an important class of monomers and they are used in many products for indoor use such as adhesives, coatings and foams. In so-called isocyanate-bonded particle boards these have completely replaced formaldehyde. Isocyanates are characterised by the electrophilic –N=C=O group, which reacts easily with OH-containing molecules such as water and alcohols. In the hydrolysis with water primary amines are formed, whereas the reaction with alcohols leads to carbamates (urethanes). Polyurethanes are obtained by polyaddition of diisocyanates and diol components. Industrially used diisocyanates essentially involve 4,4'-diphenylmethane diisocyanate (MDI), hexamethylene diisocyanate (HDI), 2,4-toluene diisocyanate and 2,6-toluene diisocyanate (2,6-TDI). Diisocyanate monomers have a strong irritant effect on the respiratory tract, eyes, skin and mucous membranes. For this reason polyisocyanates such as HDI biuret, HDI isocyanurate and poly(MDI) with low monomer proportions are mainly used in manufacturing processes. Schmidtke and Seifert [94] have investigated the HDI emissions from freshly produced polyurethane paints in test chambers. An amount of 1,800 ng m^{-3} was measured in the chamber air 2 days after the test began. An amount of 5 ng m^{-3} could still be detected after 14 days. In another study, the release of MDI from isocyanate-bonded particle board could not be determined [95]. Construction foams emit diisocyanate monomers mainly during the processing stage, as is demonstrated for 2,6-TDI in Fig. 7b [96].

4
Conclusion

Numerous investigations have shown that reactive and nonreactive substances can be introduced into an indoor environment by release from building products and indoor materials. The type and amount of emission compounds are naturally dependent on the precursor substances and the climatic parameters. Furthermore many materials commonly used indoors emit reactive compounds or secondary products owing to their manufacturing process. It has been shown that particularly secondary products can have a negative influence on people's well-being even at low concentrations owing to their odour intensity or their irritant effect. It therefore appears to be necessary to take into consideration also the aspect of indoor chemistry in the chemical characterisation

Table 6 Impact of climatic parameters on the emission rate form carpet, PVC flooring, sealant, varnish (floor) and wall paint on gypsum board. Source Wolkoff [36]

Parameter	Primary source[a]	Secondary source[b]	Comments
Air velocity $1-10$ cm s^{-1}	No significant effect after approximately 2–4 days	High air velocities may promote air oxidative emission	Evaporation-controlled emission appears to play a minor role after a few days, rather diffusion within the building products governs the emission. Secondary source emissions depend on VOC and building product type
Temperature 23–60 °C	The impact depends on building product and VOC type emitted, largest effect at 60 °C		For some products, no difference between 23 and 35 °C
Relative humidity 0–50%	The impact depends on building product and VOC type emitted	Plasticisers appear to decompose at high relative humidity	
Nitrogen/air		Some VOCs or plasticisers appear to be sensitive to air oxidation	Anaerobic testing may inform about air-susceptible building products
Repeatability	Minor differences		Any difference diminishes at long times
Homogeneity	Minor differences		

[a] Non bound VOCs in building products.
[b] Chemically bound VOCs released by both chemical and physical mechanisms.

of the indoor air and emission testing. A number of test procedures are now available for standardised evaluation of building products. However, in some cases it does not appear to be very practicable to test the emission behaviour of sensitive products and product groups generally and exclusively under standard conditions, as climatic parameters may have a strong impact on the emission rate (Table 6). Depending on the product type it is sensible to select those climatic conditions that can stimulate the possible chemical reactions and therefore the release of secondary products in order to determine the true usability of the products [39, 97]. Some of the compounds emitted may affect the perception of indoor air quality. Consequently, methods need to be developed to characterise chemical and sensory terms in order to understand their impact on perceived air quality and to assess potential health risks [40].

Appendix 1: Abbreviations

C_x: concentration of a VOC_x in the emission test chamber/cell in micrograms per cubic metre.

L: product loading factor in square metres per cubic metre.

n: air exchange rate per hour.

q: area specific air flow rate ($=n/L$) in cubic metres per square metre per hour.

SER_a: area specific emission rate in micrograms per square metre per hour.

SER_l: length specific emission rate in micrograms per metre per hour.

SER_v: volume specific emission rate in micrograms per cubic metre per hour.

SER_u: unit specific emission rate in micrograms per hour per unit.

t: time after start of the test, in hours or days.

Appendix 2: Definitions

Emission test chamber: Enclosure with controlled operational parameters for the determination of VOCs emitted from building products.

Emission test cell: Portable device for the determination of VOCs emitted from building products. The emission cell is placed against the surface of the test specimen, which thus becomes a part of the emission cell.

Air exchange rate: The ratio of the volume of clean air brought into the emission test chamber hourly and the free emission test chamber volume measured in identical units, expressed in air changes per hour.

Air flow rate: Air volume entering into the emission test chamber per unit time.

Air velocity: Air speed over the surface of the test specimen.

Area specific air flow rate: Ratio between the supply air flow rate and the area of the test specimen.

Building product: Product produced for incorporation in a permanent manner in construction works.

Emission test chamber/cell concentration: Concentration of a specific VOC_x (or group of VOCs) measured in the emission test chamber/cell outlet.

Product loading factor: Ratio of exposed surface area of the test specimen and the free emission test chamber/cell volume.

Recovery: Percentage of measured mass or a target VOC in the air leaving the emission test chamber during a given time period divided by the mass of target VOC added to the emission test chamber in the same time period.

Sample: A part or a piece of a building product which is representative of the production.

Specific emission rate: Product specific rate describing the mass of a volatile organic compound emitted from a product per unit time at a given time from the start of the test. The area-specific emission rate, SER_a, is used in the standard, Several other specific emission rates can be defined according to different requirements, for example, length-specific emission rate, SER_l, volume-specific emission rate, SER_v, and unit-specific emission rate, SER_u. The term area-specific emission rate is sometimes used in parallel with the term emission factor.

Target volatile organic compounds: Product-specific volatile organic compounds.

References

1. World Health Organization (1989) Indoor air quality: organic pollutants. EURO reports and studies 111. World Health Organization, Copenhagen
2. European Commission (1997) Total volatile organic compounds (TVOC) in indoor air quality investigations. Indoor air quality and its impact on man. Report no 19. European Commission, Luxembourg
3. Tichenor BA (1989) Indoor air sources; using small environmental test chambers to characterize organic emissions from indoor materials and products. US EPA, report no EPA-600/8-89-074
4. European Commission (1997) Evaluation of VOC emissions from building products – solid flooring materials. Indoor air quality and its impact on man. Report no 18. European Commission, Luxembourg
5. Uhde E (1998) PhD thesis. Technical University of Braunschweig
6. Sollinger S, Levsen K (1993) Atmos Environ 27B:183
7. Meininghaus R, Salthammer T, Knöppel H (1999) Atmos Environ 33:2395
8. Meyer U, Möhle K, Eyerer P, Maresch L (1994) Staub Reinhalt Luft 54:137
9. Colombo A, De Bortoli M, Knöppel H, Pecchio E, Vissers H (1993) Indoor Air 3:276
10. Jørgensen RB, Bjørseth O, Malvik B (1999) Indoor Air 9:2
11. Meininghaus R, Schauenburg H, Knöppel H (1998) Environ SciTechnol 32:1861
12. EN 13419-1 (2003) Building products – determination of the emission of volatile organic compounds, part 1: emission test chamber method. Beuth, Berlin
13. Gunnarsen L (1997) Indoor Air 7:116
14. Guo Z (1993) In: Nagda NL (ed) Modeling of indoor air quality and exposure. American Society for Testing and Materials. Philadelphia, PA, ASTM STP 1205:131
15. Wolkoff P, Clausen PA, Nielsen PA, Gunnarsen L (1993) Indoor Air 3:291
16. Wolkoff P (1996) Gefahrst Reinhalt Luft 56:151
17. Uhde E, Borgschulte A, Salthammer T (1998) Atmos Environ32:773
18. Gustafsson H (1999) In: Salthammer T (ed) Organic indoor air pollutants. Wiley-VCH, Weinheim, p 143
19. European Commission (1994) Sampling strategies for volatile organic compounds (VOCs) in indoor air. Indoor air quality and its impact on man. Report no 14. European Commission, Luxembourg
20. Wensing M, Schulze D, Salthammer T (2002) In: Moriske H-J, Turowski E (eds) Handbuch für Bioklima und Lufthygiene. ECOMED, Landsberg, p III-6.2.2

21. Salthammer T (ed) (1999) Organic indoor air pollutants. Wiley-VCH, Weinheim
22. American Society for Testing and Materials (1997) Standard guide for small-scale environmental chamber determinations of organic emissions from indoor materials/ products. ASTM D 5116–97. ASTM International, West Conhohocken, PA
23. Kephalopulos S (1999) In: Salthammer T (ed) Organic indoor air pollutants. Wiley-VCH, Weinheim, p 153
24. Salthammer T (1996) Atmos Environ 30:161
25. Bauhof H, Wensing M (1999) In: Salthammer T (ed) Organic indoor air pollutants. Wiley-VCH, Weinheim, p 105
26. Chang JCS, Tichenor BA, Guo Z, Krebs KA (1997) Indoor Air 7:241
27. Sjöberg A (1997) Floor systems, moisture and alkali. Proceedings of healthy buildings/IAQ '97, global issues and regional solutions, vol 10, p 567
28. European Commission (1993) Determination of VOCs emitted from indoor materials and products. Interlaboratory comparison of small chamber measurements. Indoor air quality and its impact on man. Report no. 13. European Commission, Luxembourg
29. European Commission (1995) Determination of VOCs emitted from indoor materials and products. Second interlaboratory comparison of small chamber measurements. Indoor air quality and its impact on man. Report no 16. European Commission, Luxembourg
30. Wensing M (1999) In: Salthammer T (ed) Organic indoor air pollutants. Wiley-VCH, Weinheim, p 129
31. European Commission (1991) Guideline for the characterization of volatile organic compounds emitted from indoor materials and products using small test chambers. Indoor air quality and its impact on man. Report no. 8. European Commission, Luxembourg
32. Nordtest (1990) Building materials: emission of volatile compounds, chamber method. NT Build 359, Finland
33. EN 13419-2 (2003) Building products – determination of the emission of volatile organic compounds, part 2: emission test cell method. Beuth, Berlin
34. EN 13419-3 (2003) Building products – determination of the emission of volatile organic compounds, part 3: procedure for sampling, storage of samples and preparation of test specimens. Beuth, Berlin
35. ISO 16000-6 (2000) Indoor air – determination of volatile organic compounds in indoor and chamber air by active sampling on Tenax TA, thermal desorption and gas chromatography MSD/FID. Beuth, Berlin
36. American Society for Testing and Materials (2001) Standard practice for full-scale chamber determination of volatile organic emissions from indoor materials/products. ASTM D 6670–01. ASTM International. West Conhohocken, PA
37. American Society for Testing and Materials (2001) ASTM standards on indoor air quality. ASTM International, West Conhohocken, PA
38. Wolkoff P, Clausen PA, Jensen B, Nielsen GD, Wilkins CK (1997) Indoor Air 7:92
39. Wolkoff P (1998) Atmos Environ 32:2659
40. Knudsen HN, Kjaer PA, Nielsen PA, Wolkoff P (1999) Atmos Environ 33:1217
41. Salthammer T, Fuhrmann F, Schwarz A (1999) Atmos Environ 33:75
42. Tucker GW (2001) In: Spengler JD, Samet JM, McCarthy JF (eds) Indoor air quality handbook. Mc Graw-Hill, New York, p 31.1
43. Salthammer T (2001) Farbe + Lack 107:130
44. Sparks LE, Guo Z, Chang JC, Tichenor BA (1999) Indoor Air 9:10
45. Sparks LE, Guo Z, Chang JC, Tichenor BA (1999) Indoor Air 9:18
46. Reiss R, Ryan PB, Koutrakis P, Tibbetts SJ (1995) Environ Sci Tech 29:1906
47. Clausen PA (1993) Indoor Air 3:269

48. Wensing M (1996) In: Aktuelle Aufgaben der Messtechnik in der Luftreinhaltung. VDI-Bericht 1257. VDI, Dusseldorf, p 405
49. Salthammer T, Bednarek M, Fuhrmann F (1999) Proceedings of Indoor Air 99, vol 5. Edinburgh, p 99
50. Salthammer T, Bednarek M, Fuhrmann F, Funaki R, Tanabe S-I (2002) J Photochem Photobiol A (in press)
51. Englund F (1999) Emissions of volatile organic compounds (VOC) from wood. Report I 9901001. Trätek, Stockholm
52. Jann O, Wilke O (1999) Proceedings Indoor Air 99, vol 5. Edinburgh, p 75
53. Lundgren B, Jonsson B, Ek-Olausson B (1999) Indoor Air 9:202
54. Brown S.K. (1999) Indoor Air 9:209
55. Kelly TJ, Myers JD, Holdren MW (1999) Indoor Air 9:117
56. Wirts M, Salthammer T (2002) Environ Sci Technol 36:1827
57. Horn W, Ullrich D, Seifert B (1998) Indoor Air 8:39
58. Salthammer T, Fuhrmann F (2000) Indoor Air 10:133
59. Saarela K (1999) In: Salthammer T (ed) Organic indoor air pollutants. Wiley-VCH, Weinheim, p 185
60. Van der Wal JF, Hoogeveen AW, Wouda P (1997) Indoor Air 7:215
61. Zellweger C, Hill M, Gehrig R, Hofer P (1997) Emissions of volatile organic compounds. EMPA, Dübendorf
62. Moriske H-J, Ebert G, Konieczny L, Menk G, Schöndube M (1998) Gesund Ing 119:90
63. Jensen B, Wolkoff P, Wilkins CK, Clausen PA (1995) Indoor Air 5:38
64. Jensen B, Wolkoff P, Wilkins CK (1995b) Indoor Air 5:44
65. Salthammer T (1997) Indoor Air 7:189
66. Salthammer T, Fuhrmann F (1996) Proceedings Indoor Air 96, vol 3. Nagoya, p 607
67. Nagda NL, Koontz MD, Kennedy PW (1993) Indoor Air 3:189
68. Black MS, Worthan AW (1999) Proceedings Indoor Air 99, vol 2. Edinburgh, p 454
69. Wensing M (1999) Proceedings Indoor Air 99, vol 5. Edinburgh, p 87
70. De Bortoli M, Kephalopoulos S, Kirchner S, Schauenburg H, Vissers H (1999) Indoor Air 9:103
71. Uhde E, Bednarek, M, Fuhrmann F, Salthammer T (2001) Indoor Air 11:150
72. Sollinger S, Levsen K (1994) Atmos Environ 28:2369
73. Knudsen HN, Clausen G, Fanger PO (1997) Indoor Air 7:107
74. Fang L, Clausen G, Fanger PO (1999) Indoor Air 9:193
75. Meininghaus R, Gunnarsen, L, Knudsen, HN (2000) Environ Sci Technol 34:3104
76. Won D, Corsi RL, Rynes M (2001) Indoor Air 11:246
77. Devos M, Patte F, Rouault J, Laffort P, Van Gemert LJ (1990) Standardized human olfactory thresholds. Oxford University Press, New York
78. Stoye D, Freitag W (1996) Lackharze. Hanser, München
79. Belitz HD, Grosch W (1992) Lehrbuch der Lebensmittelchemie. Springer, Berlin Heidelberg New York
80. Salthammer T (1999) In: Salthammer T (ed) Organic indoor air pollutants. Wiley-VCH, Weinheim, p 219
81. Breitmeier E (1999) Terpene. Teubner, Stuttgart
82. Jann O, Wilke O, Brödner D (1997) In: Woods E, Grimsrud TD, Boshi N (eds) Proceedings of healthy buildings/IAQ '97, vol 3. Washington, DC, p 593
83. Fouassier JP (1995) Photoinitiation, photopolymerisation and photocuring. Hanser, Munich
84. Clausen P, Wolkoff P (1997) Atmos Environ 31:715
85. Salthammer T, Schriever E, Marutzky R (1993) Toxicol Environ Chem 40:121

86. Thölmann D (1996) In: Proceeedings of Analytica Conference. Munich
87. Wilke O, Jann O (1999) In: Proceedings 4. Freiberger Polymertag, G/1-G/11. Forschungs-
    institut für Leder- und Kunstledertechnologie, Freiberg
88. Kawaguchi H (1994) Chemosphere 28:1489
89. WHO (1997) Flame retardants: a general introduction. Environmental health criteria,
    vol 192. World Health Organization, Geneva
90. Ingerowski G, Friedle A, Thumulla J (2001) Indoor Air 11:145
91. Pardemann J, Salthammer T, Uhde E, Wensing M (2000) In: Proceedings of healthy build-
    ings 2000, vol 4. Helsinki, p 125
92. Salthammer T, Wensing M (2002) Proceedings Indoor Air 02. vol 2. Monterey, CA, p 213
93. Salthammer T (1999) In: Salthammer T (ed) Organic indoor air pollutants. Wiley-VCH,
    Weinheim, p 203
94. Schmidtke F, Seifert B (1990) Fresenius J Anal Chem 336:647
95. Schulz M, Salthammer T (1998) Fresenius J Anal Chem 362:289
96. Salthammer T (2000) In: Moriske H-J, Turowski E (eds) Handbuch für Bioklima und
    Lufthygiene. ECOMED, Landsberg, p III-6.4.2
97. Wolkoff P (1999) Sci Total Environ 227:197

The Handbook of Environmental Chemistry Vol. 4, Part F (2004): 73–87
DOI 10.1007/b94831
© Springer-Verlag Berlin Heidelberg 2004

Adsorption and Desorption of Pollutants to and from Indoor Surfaces

Bruce A. Tichenor (✉)

IAQ Consultant, 210 Hummingbird Lane, Macon, NC 27551, USA
tichenor@gloryroad.net

Abstract Adsorption and desorption of indoor air pollutants to and from indoor surfaces are important phenomena. Often called sink effects, these processes can have a major impact on the concentration of pollutants in indoor environments and on the exposure of human occupants to indoor air pollutants. Basic theories are used to describe the processes using fundamental equations. These equations lead to models describing sink effects in indoor environments. Experimental studies have been performed to determine the important parameters of the sink models. Studies conducted in dynamic, flow-through environmental test chambers have quantified adsorption and desorption rates for many combinations of indoor air pollutants and interior surfaces. Sink effects have been incorporated into indoor air quality (IAQ) models to predict how adsorption and desorption processes affect concentrations of indoor pollutants. Limited numbers of full-scale, test house studies have been conducted to provide validation data for IAQ models in order to improve their accuracy. IAQ model predictions are useful for analyzing the interaction of indoor sources and sinks and their effect on human exposure to indoor air pollutants.

Keywords   Adsorption · Desorption · Sink effects · Sink models · Indoor air quality models

1
Introduction

Adsorption and desorption of indoor air pollutants to and from indoor surfaces are important phenomena. Often called sink effects, these processes can have a major impact on the concentration of pollutants in indoor environments and on the exposure of human occupants to indoor air pollutants. The purpose of this chapter is to discuss how sink effects impact indoor air quality (IAQ) and to highlight the published IAQ research on the subject. It is not intended to be a comprehensive literature review, so many relevant publications will not be cited. Nor are sufficient details provided to allow the conduct of complex sink evaluations. The reader is encouraged to consult the cited references for pertinent details. Also, the focus of the chapter will be on vapor-phase organic compounds; limited information is provided on other indoor pollutants.

Sorption phenomena are well described and documented in the basic scientific literature [1] and standard physical chemistry texts [2–4]. Indoor air researchers first reported extensively on adsorption to and desorption from indoor surfaces in 1987 at the 4th International Conference on Indoor Air Quality and Climate in Berlin where Seifert and Schmal [5] reported the sink effects of plywood and carpet exposed to lindane and a mixture of 20 volatile organic compounds (VOCs). The Berlin conference also provided data from two studies that examined the sink effect of textiles [6, 7]. Finally, Skov and Valbjorn [8] reported on the seminal Danish town hall study. They concluded that sorption to and from large surface areas and fleecy materials was associated with IAQ problems. This work was followed up by Nielsen [9] at Healthy Buildings '88 in Stockholm, where he reviewed several studies on the importance of sorption processes on IAQ. Also in Stockholm, Berglund and her colleagues [10] provided an extensive literature review and discussion of IAQ data leading to the conclusion that sink effects are important. They also provided the results of a study on adsorption and desorption in a ventilation system. Since then, numerous overviews have been presented on the importance of sink effects on IAQ [11, 12].

1.1
Vapor-Phase Organic Compounds

The majority of research on IAQ sink effects has focused on vapor-phase organic compounds. These gases are often divided into two categories: VOCs and semivolatile organic compounds (SVOCs). Much of the remainder of this chapter will deal with sorption phenomena associated with VOCs and SVOCs. However, other indoor pollutants that interact with indoor surfaces are also important.

1.2
Other Indoor Pollutants

Among the non-vapor-phase organic pollutants studied for indoor sink effects are NO_2, ozone, and particles. Some of the earliest work on indoor sink effects was associated with combustion source emissions, namely NO_2. In 1989, Spicer et al. [13] reported on the rates and mechanisms of NO_2 removal from indoor air by residential materials. Later, work by Weschler and colleagues [14] showed how indoor ozone reacts with NO_2 to form nitric acid, which adsorbs on indoor surfaces. Additional work by Weschler and colleagues [15] reported on rates of ozone and NO_2 adsorption on indoor surfaces using deposition velocities and how the adsorbed NO_2 can react to form NO and HONO. Cano-Ruiz and Nazaroff [16] developed a theoretical model of the removal of ozone at indoor surfaces by combining mass transport and surface kinetics. Later, Morrison and Nazaroff [17] studied the removal of ozone by ventilation duct material using models based on deposition velocities. They concluded that ventilation ducts are probably not important ozone sinks. Indoor particles can also sorb on surfaces. Recent work by Lai and Nazaroff [18] describes how deposition rates can vary by orders of magnitude for typical indoor turbulence levels. An intriguing study of particle/gas interaction was conducted by Johansson [19], who reported that environmental tobacco smoke adsorbed on indoor surfaces can undergo gas-to-particle conversion and re-emit particles.

2
Adsorption/Desorption Processes

Basic theories are used to describe the processes using fundamental equations.

2.1
Theory

The interaction of VOC (and SVOC) molecules with indoor surfaces can be described by well-documented adsorption theories [3]. Three isotherm theories are available to quantify indoor air sink effects: Langmuir, Freundlich, and

Brunauer–Emmett–Teller (BET) [20]. Langmuir isotherm theory is applied when a monolayer of molecules is sorbed on a homogeneous surface where each site requires the same energy to adsorb molecules. The Freundlich isotherm also assumes a monolayer, but allows for an exponential distribution of adsorption energies. BET isotherms provide for multilayer adsorption.

2.2
Fundamental Equations

For indoor organic gases (VOCs and SVOCs) Langmuir isotherm theory is commonly used to describe indoor sink behavior at equilibrium [20]:

$$k_a C_e (1 - \Theta) = k'_d \Theta, \tag{1}$$

where k_a is the adsorption rate constant, C_e is the equilibrium concentration of the adsorbate in the vapor phase, Θ is the proportion of available adsorption sites that are occupied, and k'_d is a desorption constant. For the low values of C_e encountered in indoor air it can be assumed that the occupied sites are a very small proportion of the available sites, so Eq. (1) can be rewritten as

$$k_a C_e = k'_d \Theta. \tag{2}$$

The equilibrium mass in the sink (M_e) is proportional to Θ, so Eq. (2) leads to

$$k_a C_e = k_d M_e. \tag{3}$$

If a Freundlich isotherm is assumed,

$$M = kC^a, \tag{4}$$

where M is the mass in the sink, C is the gas-phase concentration, and k and a are empirical constants.

3
Models

The previous equations describing the adsorption/desorption behavior of gases lead to models describing sink effects in indoor environments. In addition, transport of molecules within the sink material can have a major impact on desorption rates; thus, models accounting for internal diffusion have been developed for indoor sinks. Models based on fundamental theories are preferred over empirical approaches, but some studies rely on experimental data to fit empirical models [21–23].

3.1
Adsorption Models

Using conventional units, and assuming a sink area of A, Eq. (3) can be written as an adsorption/desorption mass balance model:

$$k_a C_e A = k_d M_e A,$$ (5)

where k_a is the adsorption rate (micrograms per hour), C_e is the equilibrium VOC (or SVOC) concentration (micrograms per cubic meter), A is the sink area (square meters), k_d is the desorption rate constant (per hour), and M_e is the equilibrium sink mass (micrograms per square meter). In addition, an equilibrium constant can be defined as

$$k_e = k_a/k_d = M_e/C_e,$$ (6)

where k_e is the equilibrium constant (meters). k_e is sometimes called the partition coefficient. For a given combination of VOC and sink material, k_e describes the sink strength; the larger the value of k_e, the greater the mass adsorbed on the sink at equilibrium. Note that k_e has been mistakenly used to denote sink capacity. The capacity of an indoor sink would be the maximum mass adsorbed under extremely high indoor concentrations where the assumptions of the Langmuir isotherms do not apply and Eq. (6) is invalid.

3.2
Diffusion Models

As noted by several authors [24–29], sink models based solely on adsorption theories fail to account for other important phenomena. Axley [24] provided equations for diffusion processes, both within the boundary layer adjoining the sinks and within the sinks themselves, which describe the behavior of indoor sinks. Little and Hodgson [26] presented a strategy for characterizing homogenous, diffusion-controlled sinks using heat transfer analogies. Jørgensen et al. [27] presented a sink diffusion model that couples internal diffusion with surface adsorption. They showed that Ono–Kondo Langmuir models are sufficient to describe weak sorption situations, but internal diffusion models are needed when stronger sorption occurs. Internal diffusion of a specific VOC in a specific sink material can be represented by Fick's law as [27]

$$F(x) = -D(dC/dx),$$ (7)

where $F(x)$ is the mass flux of the VOC at a distance x from the surface of the sink material (micrograms per square meter per hour), D is the diffusion coefficient of the VOC in the sink material (square meters per hour), and dC/dx is the concentration gradient in the sink (micrograms per meter to the fourth power).

3.3
Properties Affecting Sink Behavior

The previous equations represent the most fundamental processes that describe the behavior of indoor sinks. As such, they provide insight into what properties of the VOCs and/or the sink are important. For example, as already noted, k_e, the equilibrium constant (or partition coefficient), is proportional to the sink strength. Thus, for a specific sink material, an increase in k_e means an increase in sink strength. As shown by several studies, k_e can be related to a physical property of a VOC. Colombo et al. [30] reported an increase in k_e with an increase in compound boiling point, while Kjaer et al. [31] showed a similar relationship with gas chromatography retention time. Finally, Corsi et al. [32] and Weschler [33] presented a correlation between the equilibrium constant and a compound's saturation vapor pressure showing the same relationship. Thus, as a general rule, the lower the compound's volatility, the stronger the sink effect. Environmental factors that affect the compound or sink properties can also be important. For example, Guo et al. [34] presented an analysis of perchloroethylene adsorption on fabrics showing how higher temperatures increase desorption rates. Finally, of course, the nature of the sink material is important. Rough, textured surfaces (e.g., carpet) and adsorbent surfaces which promote internal diffusion (e.g., gypsum board) are stronger sinks than smooth, dense material (e.g., glass.).

4
Experimental Studies

In order to use sink models, important parameters must be available. For example, the Langmuir adsorption model requires information on the rates of adsorption and desorption. Diffusion models require information on the diffusion coefficients. These parameters are dependent upon the characteristics of both the VOC (or SVOC) and the sink material, and fundamental data are generally not available. Thus, experimental studies are required to determine the values of the important parameters of the sink models.

4.1
Test Methods and Protocols

A number of test methods have been used to determine sink model parameters. The most common test protocol uses a dynamic, flow-through chamber and involves challenging a test sink material with a test gas [20, 31, 35, 36]. Details on this technique are presented later. Other methods include static tests and microbalance measurements. Borrazzo et al. [37] took a fundamental physical chemistry approach and used static equilibrium tests to determine partition coefficients for trichloroethylene and ethanol vapors and several types of fibers. Two studies reported on the use of microbalances to determine the mass adsorbed and desorbed and the resultant adsorption and desorption rates [26, 38].

4.2
Chamber Studies

The use of dynamic, flow-through test chambers is common in the study of emissions from sources of indoor air pollution [39]. They are also widely used in the study of indoor sinks. Some of the earlier work on sinks examined the surfaces of the test chambers themselves and showed that chamber sink effects can be important [21]. Researchers routinely evaluate test chamber systems for sink effects [34, 40]. A recent paper on the measurement of SVOC emissions showed how the chamber sink effect can be exploited [41]. In this study, SVOCs adsorbed on chamber walls were removed by heating, flushed out, and quantified to give SVOC emission rates.

The most common method of evaluating indoor sinks involves placing a sample of the sink material in a chamber and then flowing a fixed concentration of challenge gas through the chamber at a known flow rate. This is shown schematically in Fig. 1.

For the sink test shown in Fig. 1 and using mass-balance equations developed previously [21], Tichenor et al. [20] used Langmuir's adsorption theory to develop the following equations that describe how the VOC (or SVOC) concentration in the chamber and the mass in the sink change over time:

$$dC/dt = NC_{in} - NC - k_aCL + k_dML, \tag{8}$$

where N is the chamber air change rate (per hour), C_{in} is the inlet concentration (micrograms per cubic meter), C is the chamber and outlet concentration at any time (micrograms per cubic meter), $L = A/V$ (per meter), and k_a and k_d are as previously defined.

$$dM/dt = k_aC - k_dM, \tag{9}$$

where M is the mass per unit area in the sink at any time. For a typical sink test, a sample of sink material (area A) is placed in a test chamber (volume V) and

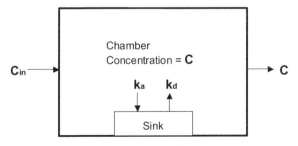

Fig. 1 Flow-through chamber sink test

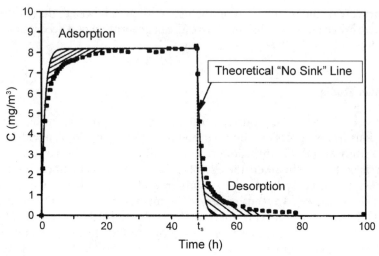

Fig. 2 Chamber sink test data

at time zero, the flow (Q) is started with an incoming concentration of C_{in}. The concentration in the chamber (C) is monitored and when it reaches equilibrium (C_e) the challenge gas is turned off (C_{in}=0) at time t_s. The time between 0 and t_s is the adsorption phase of the test, and the time after t_s is the desorption phase. Figure 2 shows data from a typical sink test, as well as the "no-sink" line [20].

On the basis of Eqs. (8) and (9), there are three unknown values: M, k_a, and k_d. M=M_e at time t_s and is determined by the cross-hatched area in Fig. 2. k_a and k_d are determined by nonlinear regression curve fits [20]. Figure 3 shows data and a fitted Langmuir curve for the desorption portion of the test, along with the "no-sink" line [20].

Figures 1, 2, and 3 are provided to illustrate one protocol often used to evaluate sink materials [20, 32, 42–47]; however, other methods are also used. For example, Krebs and Guo [48] reported on a unique method involving two test chambers in series. The first chamber is injected with a known concentration of a pollutant (in this case, ethylbenzene). The outlet from the first chamber provides a simple first-order decay that is injected into the inlet of the second chamber that contains the sink material (gypsum board). Thus, this method exposes the sink test material to a changing concentration typical of many wet VOC sources. The sink adsorption rate and desorption rate results are comparable to one-chamber tests and are achieved in a much shorter experimental time. Kjaer et al. [31] reported on using a CLIMPAC chamber and sensory evaluations coupled with gas chromatography retention times to evaluate desorption rates. Finally, Funaki et al. [49] used ADPAC chambers and exposed sink materials to known concentrations of formaldehyde and toluene and then desorbed the sinks using clean air. They reported adsorption rates as a percentage of concentration differences.

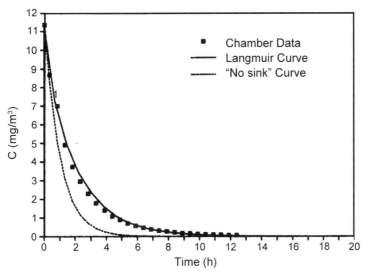

Fig. 3 Data and fitted Langmuir curve

4.3
Full-Scale Studies

A limited number of sink effect studies have been conducted in full-sized environments. Tichenor et al. [20] showed the effect of sinks on indoor concentrations of total VOCs in a test house from the use of a wood stain. Sparks et al. [50] reported on test house studies of several indoor VOC sources (i.e., p-dichlorobenzene moth cakes, clothes dry-cleaned with perchloroethylene, and aerosol perchloroethylene spot remover) and they were compared with computer model simulations. These test house studies indicated that small-chamber-derived sink parameters (k_a and k_d) may not be applicable to full-scale, complex environments. The re-emission rate (k_d) appeared to be much slower in the test house. This result was also reported by other investigators in a later study [51]. New estimates of k_a and k_d were provided, including estimates of k_a (or deposition velocity) based on the diffusivity of the VOC molecule [50]. In a test house study reported by Guo et al. [52], ethylbenzene vapor was injected at a constant rate for 72 h to load the sinks. Re-emissions from the sinks were determined over a 50-day period using a mass-balance approach. When compared with concentrations that would have occurred by simple dilution without sinks, the indoor concentrations of ethylbenzene were almost 300 times higher after 2 days and 7 times higher after 50 days. Studies of building bake-out have also included sink evaluations. Offermann et al. [53] reported that formaldehyde and VOC levels were reduced only temporarily by bake-out. They hypothesized that the sinks were depleted by the bake-out and then returned to equilibrium after the post-bake-out ventilation period. Finally, a test house study of latex paint emissions and sink effects again showed that small-chamber-derived sink parameters for various latex paint components do not agree with those determined using test house data [54].

5
Experimental Results

Studies conducted in dynamic, flow-through environmental test chambers have quantified adsorption and desorption rates for only a small number of the available combinations of indoor air pollutants and interior surfaces.

5.1
Adsorption/Desorption Rates

Experimental studies to determine adsorption and desorption rates show wide differences in values depending on the sink material and the VOC adsorbate. Because of the large number of possible combinations of indoor surfaces and VOCs, only a few of these combinations have been evaluated. References that give values of k_a and k_d for a variety of indoor sink surfaces and indoor pollutants based on dynamic small chamber testing are provided in Table 1.

A review of the references cited in Table 1 would show a wide difference in k_a and k_d values depending on the sink material and pollutant. This is illustrated in Table 2, where a few representative values are provided.

Table 1 Sink materials and indoor pollutants evaluated for k_a and k_d

Reference	Sink material	Indoor pollutant
[42]	Carpet	Toluene, α-pinene
[55]	Dust	2-Butoxyethanol
[32]	Carpet, carpet pad, painted and unpainted gypsum board, wallpaper, vinyl flooring, ceiling tile	Methyl tert-butyl ether, cyclohexane, isopropyl alcohol, toluene, tetrachloroethylene, ethylbenzene, o-dichlorobenzene, 1,2,4-trichlorobenzene
[20]	Carpet, gypsum board, ceiling tile, window glass, pillow	Ethylbenzene, tetrachloroethylene
[27]	Carpet, PVC flooring	Toluene, α-pinene
[46]	Carpet, gypsumboard	Ethylene glycol, propylene glycol, butoxyethoxyethanol, Texanol (all are latex paint components)
[56]	Paint (on aluminum sheet), painted gypsum board	Toluene, tetrachloroethylene, isopropyl alcohol, cyclohexane, methyl ethyl ketone, ethyl acetate

Table 2 Estimates of k_a and k_d from small chamber tests

Reference	Sink material	Indoor pollutant	k_a (m h^{-1})	k_d (h^{-1})	k_e (m)
[32]	Carpet	Tetrachloroethylene	0.31	0.32	0.97
[32]	Carpet	Ethylbenzene	0.41	0.34	1.21
[32]	Gypsum board	Ethylbenzene	0.07	0.27	0.26
[32]	Vinyl flooring	Ethylbenzene	0.06	0.22	0.27
[20]	Carpet	Tetrachloroethylene	0.13	0.13	0.97
[20]	Carpet	Ethylbenzene	0.08	0.08	0.95
[20]	Gypsum board	Ethylbenzene	0.45	1.5	0.30
[56]	Gypsum board	Toluene	0.08	0.35	0.22
[27]	Carpet	Toluene	0.25	0.41	0.62
[27]	PVC flooring	Toluene	0.065	0.26	0.25
[46]	Carpet	Texanol	0.84	0.016	52
[46]	Gypsum board	Texanol	1.76	0.0048	370

Table 2, while showing only a few values, can be used to illustrate several concepts:

- Studies of the same sink/VOC combination show wide variations in k_a and k_d, but similar values for k_e.
- For most pollutants, fleecy materials (e.g., carpet) are stronger sinks (i.e., higher k_e) than smooth materials (e.g., vinyl and PVC flooring, gypsum board)
- VOCs in latex paint (e.g., Texanol) desorb very, very slowly.

5.2
Model Validation

Limited numbers of full-scale, test house studies have been conducted to provide validation data for IAQ models in order to improve their accuracy. Estimates of k_a and k_d based on test house studies are provided in Table 3.

In all cases, when the values for k_a and k_d obtained from full-scale studies are compared with small chamber studies, the desorption rates (k_d) are much slower in the full-scale environments. Thus, sink parameters obtained from small chamber studies must be used with great caution.

The difference between chamber-derived Langmuir estimates of the sink effect and test house results using a Freundlich-type model, denoted as Best Fit [20], is illustrated in Figure 4. Note how the Langmuir model does not provide for the long term re-emissions from the sink.

Table 3 Estimates of k_a and k_d from testhouse studies

Reference	Sink material	Indoor pollutant	k_a (m h^{-1})	k_d (h^{-1})	k_e (m)
[50]	Carpet	General volatile organic compounds and tetra-chloroethylene	0.1	0.008	12.5
[50]	Carpet	p-Dichlorobenzene	0.2	0.008	25
[50]	Painted walls and ceilings	General volatile organic compounds and tetra-chloroethylene	0.1	0.1	1
[50]	Painted walls and ceilings	p-Dichlorobenzene	0.2	0.008	25
[54]	Test house surfaces	Ethylene glycol	3.2	0.0001	32,000
[54]	Test house surfaces	Propylene glycol	3.2	0	∞
[54]	Test house surfaces	Butoxyethoxyethanol	1.1	0.003	370
[54]	Test house surfaces	Texanol	1	0.0002	5,000

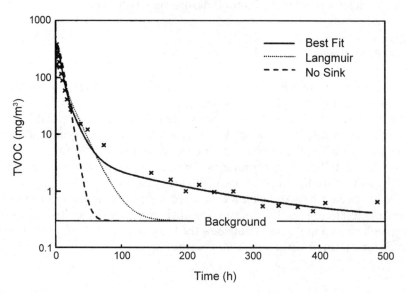

Fig. 4 Concentration of total volatile organic compounds from a wood stain applied in a test house

6
Indoor Air Quality Model Studies

IAQ model predictions are useful for analyzing the interaction of indoor sources and sinks and their effect on human exposure to indoor air pollutants. Sink effects have been incorporated into IAQ models to predict how adsorption and desorption processes affect concentrations of indoor pollutants [20, 24, 50, 54, 57–60]. For example, sink/pollutant combinations with slow adsorption rates (low k_a) and fast desorption rates (high k_d) have little impact on exposures. On the other hand, sink/pollutant combinations with fast adsorption rates (high k_a) and slow desorption rates (low k_d) can slightly lower the maximum concentration of indoor pollutants, but lengthen the time of exposure to low concentrations of the pollutant. For some pollutants (e.g., latex paint components), the exposure can be extended by months or years owing to re-emissions from the sinks.

7
Conclusions

The theories describing adsorption and desorption phenomena are well established. Unfortunately, only limited studies have been performed to develop parameters applicable to indoor environments. Small chamber data are usually insufficient to describe the sink behavior (especially desorption rates) in full-scale situations, and only a very small number of test house studies have been conducted to develop sink parameters. Sink effects can have a major impact on the long-term concentrations of pollutants in indoor environments and on the exposure of human occupants to indoor air pollutants. While IAQ models can be used to estimate the effect of sinks on exposures, such estimates can be improved when better data are available on sinks. Additional research is needed to fully understand and describe the behavior of sinks in indoor environments.

References

1. DeBoer JH (1968) The dynamical character of adsorption, 2nd edn. Oxford University Press, London
2. Adamson AW (1976) Physical chemistry of surfaces, 3rd edn. Wiley, New York
3. Daniels F, Alberty R (1961) Physical chemistry, 2nd edn. Wiley, New York
4. Levine IN (1978) Physical chemistry. McGraw-Hill, New York
5. Seifert B, Schmal HJ (1987) In: Seifert B, Esdorn M, Fischer M, Rüden H, Wegner J (eds) Indoor air '87, vol 1. Institute for Water, Soil, and Air Hygiene, Berlin, p 252
6. Korte F, Gebegufi I (1987) In: Seifert B, Esdorn M, Fischer M, Rüden H, Wegner J (eds) Indoor air '87, vol 1. Institute for Water, Soil, and Air Hygiene, Berlin, p 239
7. Zsolnay A, Gebefugi I, Korte F (1987) In: Seifert B, Esdorn M, Fischer M, Rüden H, Wegner J (eds) Indoor air '87, vol 1. Institute for Water, Soil, and Air Hygiene, Berlin, p 265

8. Skov P, Valbjorn O (1987) In: Seifert B, Esdorn M, Fischer M, Rüden H, Wegner J (eds) Indoor air '87, vol 1. Institute for Water, Soil, and Air Hygiene, Berlin, p 439
9. Nielsen P (1988) In: Berglund B, Lindvall T (eds) Healthy buildings '88, vol 3. Swedish Council for Building Research, Stockholm, p 391
10. Berglund B, Johansson I, Lindvall T (1988) In: Berglund B, Lindvall T (eds) Healthy buildings '88, vol 3. Swedish Council for Building Research, Stockholm, p 299
11. Tichenor BA (1992) Ann NY Acad Sci 641:63
12. Tichenor BA (1996) In: Tichenor BA (ed) Characterizing sources of indoor air pollution and related sink effects. ASTM, West Conshohoken, PA, p 9
13. Spicer CW, Coutant RW, Ward GF, Joseph DW, Gaynor AJ, Billick IH (1989) Environ Int 15:643
14. Weschler CJ, Brauer M, Koutrakis P (1992) Environ Sci Technol 26:179
15. Weschler CJ, Shields HC, Nalk DV (1994) Environ Sci Technol 28:2120
16. Cano-Ruiz JA, Nazaroff WW (1993) In: Saarela K, Kalliokoski P, Seppänen O (eds) Indoor air '93, vol 2, Helsinki, p 555
17. Morrison GC, Nazaroff WW (1997) In: Engineering solutions to indoor air quality problems. A&WMA/EPA, Pittsburgh, PA, VIP-75, p 514
18. Lai ACK, Nazaroff WW (2000) J Aerosol Sci 31:463
19. Johansson, J (1996) In: Tichenor BA (ed) Characterizing sources of indoor air pollution and related sink effects. ASTM, West Conshohoken, PA, p 134
20. Tichenor BA, Guo Z, Dunn JE, Sparks LE, Mason MA (1991) Indoor Air 1:23
21. Dunn JE, Tichenor BA (1988) Atmos Environ 22:885
22. Guo Z, Dunn JE, Tichenor BA, Mason MA, Krebs KA (1990) In: Proceedings of the 5th international conference on indoor air quality and climate, vol 4, Toronto, p 177
23. Sambinello D, Piva S (2002) In: Levin H (ed) Indoor air 2002: Proceedings of the 9th international conference on indoor air quality and climate, vol 3, Santa Cruz, CA, p 552
24. Axley JW (1991) Indoor Air 2:147
25. Dunn JE, Chen T (1993) In: Nagda NL (ed) Modeling of indoor air quality and exposure. ASTM, Philadelphia, PA, p 64
26. Little, JC, Hodgson, AT (1996) In: Tichenor BA (ed) Characterizing sources of indoor air pollution and related sink effects. ASTM, West Conshohoken, PA, p 294
27. Jørgensen RB, Dokka TH, Bjørseth O (2000) Indoor Air 10:27
28. Hansson P, Stymne H (2002) In: Levin H (ed) Indoor air 2002: Proceedings of the 9th international conference on indoor air quality and climate, vol 3, Santa Cruz, CA, p 546
29. Damian A, Blondeau P, Tiffonnet AL (2002) In: Levin H (ed) Indoor air 2002: Proceedings of the 9th international conference on indoor air quality and climate, vol 3, Santa Cruz, CA, p 664
30. Colombo A, DeBortoli M, Knöppel H, Pecchio E, Vissers H (1993) In: Saarela K, Kalliokoski P, Seppänen O (eds) Indoor air '93, vol 2, Helsinki, p 407
31. Kjaer UD, Nielsen, PA, Vejrup, KV, Wolkoff, P (1996) In: Tichenor BA (ed) Characterizing sources of indoor air pollution and related sink effects. ASTM, West Conshohoken, PA, p 123
32. Corsi RL, Won D, Rynes M (1999) In: Proceedings of the 1st NSF international conference on indoor air health. NSF International, Ann Arbor, MI, p 250
33. Weschler CJ (2002) In: Levin H (ed) Indoor air 2002: Proceedings of the 9th international conference on indoor air quality and climate, vol 1, Santa Cruz, CA, p 1
34. Guo Z, Tichenor BA, Mason MA, Plunket CM (1990) Environ Res 52:107
35. DeBortoli M, Knöppel H, Columbo A, Kefalopoulos S (1996) In: Tichenor BA (ed) Characterizing sources of indoor air pollution and related sink effects. ASTM, West Conshohoken, PA, p 305

36. van der Wal JF, Hoogeveen AW, van Leeuwen L (1998) Indoor Air 8:103
37. Borrazzo JE, Davidson CI, Andelman JB (1993) In: Nagda NL (ed) Modeling of indoor air quality and exposure. ASTM, Philadelphia, PA, p 25
38. Kjaer UD, Nielsen PA (1991) In: IAQ'91 healthy buildings. ASHRAE, Atlanta, GA, p 285
39. ASTM (1998) Standard guide for small-scale environmental determinations of organic emissions from indoor materials/products, D 5116-97. ASTM, West Conshohocken, PA
40. Mason MA, Howard EM, Guo Z, Bero M (1997) In: Engineering solutions to indoor air quality problems. A&WMA/EPA, Pittsburgh, PA, VIP-75, p 45
41. Hoshino K, Imanaka T, Iwasaki T, Kato S (2002) In: Levin H (ed) Indoor air 2002: Proceedings of the 9th international conference on indoor air quality and climate, vol 2, Santa Cruz, CA, p 950
42. Jorgensen RB, Knudsen HN, Fanger PO (1993) In: Saarela K, Kalliokoski P, Seppänen O (eds) Indoor air '93, vol 2, Helsinki, p 383
43. Levsen K, Sollinger S (1993) In: Saarela K, Kalliokoski P, Seppänen O (eds) Indoor air '93, vol 2, Helsinki, p 395
44. DeBortoli M, Knöppel H, Columbo A, Kefalopoulos S (1996) In: Tichenor BA (ed) Characterizing sources of indoor air pollution and related sink effects. ASTM, West - Conshohoken, PA, p 305
45. Jørgensen RB, Bjørseth O, Malvik B (1999) Indoor Air 9:2
46. Sparks LE, Guo Z, Chang JC, Tichenor BA (1999) Indoor Air 9:11
47. Won D, Corsi RL, Rynes M (2001) Indoor Air 11:246
48. Krebs K, Guo Z (1992) In: Proceedings of the 1992 US EPA/A&WMA international symposium on measurement of toxic and related air pollutants, Pittsburgh, PA, VIP-25, p 45
49. Funaki R, Tanaka H, Tanabe S (2002) In: Levin H (ed) Indoor air 2002: Proceedings of the 9th international conference on indoor air quality and climate, vol 3, Santa Cruz, CA, p 540
50. Sparks LE, Tichenor BA, White JB, and Jackson MD (1991) Indoor Air 1:577
51. Antonelli L, Mapelli E, Strini A, Cerulli T, Leoni R, Stella S (2002) In: Levin H (ed) Indoor air 2002: Proceedings of the 9th international conference on indoor air quality and climate, vol 2, Santa Cruz, CA, p 584
52. Guo Z, Mason MA, Gunn, KN, Krebs KA, Moore SA, Chang, JCS (1992) In: Proceedings of the 1992 US EPA/A&WMA international symposium on measurement of toxic and related air pollutants, Pittsburgh, PA, VIP-25, p 51
53. Offermann FJ, Loiselle, SA, Ander, GD, Lau H (1993) In: Seppänen O, Ilmarinen R, Jaakkola JK, Kukkonen E, Säteri J, Vuorelma H (eds) Indoor air '93, vol 6, Helsinki, p 687
54. Sparks LE, Guo Z, Chang JC, Tichenor BA (1999) Indoor Air 9:18
55. Kjaer UD, Nielsen PA (1993) In: Saarela K, Kalliokoski P, Seppänen O (eds) Indoor air '93, vol 2, Helsinki, p 579
56. Popa J, Haghighat F (2002) In: Levin H (ed) Indoor air 2002: Proceedings of the 9th international conference on indoor air quality and climate, vol 3, Santa Cruz, CA, p 558
57. Sparks LE, Tichenor BA, White JB (1993) In: Nagda NL (ed) Modeling of indoor air quality and exposure. ASTM, Philadelphia, PA, p 245
58. Jayjock MA, Doshi, DR, Nungesser, EH, Shade, WD (1995) Am Ind Hyg Assoc J 56:546
59. Tichenor BA, Sparks LE (1996) Indoor Air 6:259
60. Sparks LE (2002) In: Levin H (ed) Indoor air 2002: Proceedings of the 9th international conference on indoor air quality and climate, vol 3, Santa Cruz, CA, p 274

The Handbook of Environmental Chemistry Vol. 4, Part F (2004): 89–116
DOI 10.1007/b94832
© Springer-Verlag Berlin Heidelberg 2004

Sources and Impacts of Pesticides in Indoor Environments

Werner Butte (✉)

Faculty of Chemistry, University of Oldenburg, PO Box 2503, 26111 Oldenburg, Germany
werner.butte@uni-oldenburg.de

Abstract Indoor contamination is one source of exposure to toxic pollutants and has been classified as a high environmental risk. Epidemiological research linked health effects including childhood leukemia and neuroblastoma to the indoor occurrence of pesticides. Pesticides in indoor environments contribute to human exposure via inhalation, nondietary ingestion and dermal contact. Sources for pesticides indoors are direct applications, pesticides used in varnishes, colors, adhesives, etc., or in finishing textiles, leather, carpets, etc., and pesticides brought in from outdoors. Results for pesticides in indoor environments from different countries and obtained under different conditions are compiled in this chapter. They are discussed by applying two approaches: (1) the comparison with reference values and (2) the application of a risk assessment methodology related to the hazard of the compound, leading to standards or guideline values.

Keywords   Indoor air · Indoor environment · House dust · Pesticides · Exposure

1
Introduction

On average, people in moderate climates are assumed to spend up to 95% of their time indoors and most of this time is spent in their homes [1]. Residents of the Federal Republic of Germany, depending on season and vocational activity, spend between 80% and 90% of their time indoors [2]. The National Human Activity Pattern Survey of the USA recorded adults spent an average of 87% of their time in enclosed buildings and about 6% of their time in enclosed vehicles [3].

During the last 2 decades there has been increasing concern over the effects of indoor contamination on health. Changes in building design intended to improve energy efficiency have meant that modern homes are frequently more airtight than older structures [4].

We know much less about the health risks from indoor air pollution than we do about those attributable to the contamination of outdoor air [4]. Several studies have shown that for inhabitants, especially children and other vulnerable subgroups, the home environment may be a dominant source of exposure to toxicants, including pesticides [5]. Indoor pollution has been ranked by the United States Environmental Protection Agency Advisory Board (US-EPA) and the Centers for Disease Control as a high environmental risk [6].

A classification of organic indoor contaminants, according to their volatility, was given by a WHO working group on organic indoor air pollutants [7]. This group initiated the common practice of dividing organic chemicals according to boiling points (Table 1).

Very volatile organic compounds and volatile organic compounds are transitory and predominantly found in air; pesticides and other organic compounds with a low volatility or high polarity are either semivolatile organic compounds (SVOCs) or particulate organic matter (POM).

Pesticides are either semivolatile or "nonvolatile". For example, the boiling points of DDT (260 °C), lindane (around 320 °C) and pentachlorophenol (PCP) (around 310 °C) [8] classify these pesticides as semivolatile. SVOCs partition between air and house dust, whereas POM are exclusively found in house dust. Methods to analyze pesticides in indoor air (semivolatile pesticides) and in house dust (semivolatile and particle-bound pesticides as well as particulate pesticides), sources for their occurrence indoors, concentrations found in indoor environments as well as impacts are reviewed in this chapter.

Table 1 Classification of organic indoor pollutants (after Ref [7])

Abbreviation	Description	Boiling point range (°C)	Examples
VVOC	Very volatile (gaseous) organic compounds	<0 to 50–100	Carbon monoxide, carbon dioxide, formaldehyde
VOC	Volatile organic compounds	50–100 to 240–260	Solvents (aliphatic, aromatic), terpenes
SVOC	Semivolatile organic compounds	240–260 to 380–400	Pesticides (e.g., chlorpyrifos, lindane, pentachlorophenol) plasticizers (e.g., phthalates)
POM	Particulate organic matter	>380	Pesticides (e.g., pyrethroids), polycyclic aromatic hydrocarbons

2
Sources for Pesticides Indoors

Indoor pesticides can emanate from a range of sources. They include (1) pesticide application indoors (insecticides, termiticides, wood-protecting agents, etc.), (2) tracking-in of pesticides from outdoors, and (3) impregnated textiles and carpets, treated wood, etc. from an indoor environment itself.

2.1
Pesticide Application

Pesticides are often applied in homes and buildings to protect them from fungi or to fight flying or crawling pests (cockroaches, fleas, flies, mosquitoes, termites, etc.) either by pest-control workers or by the occupants themselves. Most of the compounds used are identical to those in formulations for plant protection or against forest and agricultural pests [9]. In many countries non-commercial preparations for indoor use do not need any official registration by authorities. Data collected from 238 families in Missouri (USA) during telephone interviews showed that nearly all families (97.8%) used pesticides at least once and more than two thirds used pesticides more than five times per year [10]. From in-home interviews and inventories performed as part of the National Human Exposure Assessment Survey (NHEXAS) it was reported that pesticide products were found in 97% and used in 88% of the households in the study [11]. An assessment of the quality of the self-reported pesticide product use history suggests that participants provide plausible information regarding their pesticide use [12].

Concentrations of pesticides in indoor environments are dependent on their usage. Dust samples, for example, collected using vacuum cleaners, in households where wood preservatives were applied showed a median value of 13 mg g^{-1} of PCP, as compared with 0.008 mg g^{-1} in samples from the control group [13].

On treating cockroaches and termites high exposures to chlorpyrifos in indoor air were observed, especially for the applicators [14]. Elevated concentrations of chlorpyrifos in the air (and soil) of houses were detectable up to 8 years after its application [15–17].

2.2
Tracking-In

Pesticides may be tracked into the indoor environment after a certain outdoor application [18, 19] or through transport from the workplace to the home (para-occupational or take-home exposure). Collection of floor dust both prior to and after lawn-applied 2,4-dichlorophenoxyacetic acid (2,4-D), indicated that turf residues are transported indoors [20]. Carpet dust levels of 2,4-D and dicamba and carpet surface dislodgeable residue levels were highly correlated with turf dislodgeable residue levels [21].

Foot traffic was a significant mechanism for transport of lawn-applied pesticides (2,4-D) into homes [19, 20]. Carpet dust collected from homes having high child and pet activities had greater levels of pesticide residues than homes with low child and pet activities. Pets were identified to assist children with access to soil by digging or by accumulating soil and dust in their fur [20, 22]. Re-suspension of floor dust was the major source of 2,4-D in indoor air; re-suspended floor dust was also a major source of 2,4-D on tables and window-sills [23].

Carpets appear to act as a main storage reservoir for some of the persistent contaminants [24–26] as carpeted areas typically have a higher dust mass per unit area ("load") than uncarpeted areas [27].

Pesticides may further be transported from the workplace to the home on clothes, shoes, hair and skin [28, 29]. Household dust concentrations of azino-phosmethyl, chlorpyrifos, parathion and phosmet were significantly lower in reference homes when compared with farmer/farmworker homes [30]. Organophosphate pesticides in house dust were elevated in homes of agrarian families (household members engaged in agricultural production) when compared with nonagricultural reference homes in the Seattle (USA) metropolitan area. Dialkyl phosphate metabolites measured in children's urine were also elevated for the agrarian children [31].

2.3
Treated Indoor Furnishing

A source often unknown for indoor contamination with pesticides is treated material; for example, permethrin used to impregnate carpets and textiles

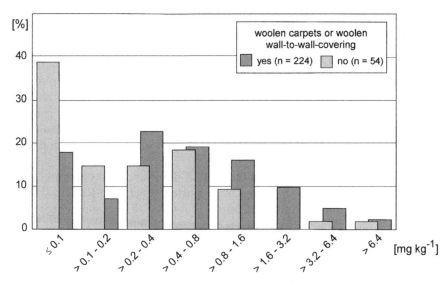

Fig. 1 Residues of DDT house dust in homes with and without woolen carpets or woolen wall-to-wall covering

and PCP evaporating from treated timber. In many cases consumers are not aware of the fact that upon buying a new woolen carpet they are bringing, for example, permethrin or DDT indoors that might disperse in the indoor environment. Treated material may be regarded as a secondary source of exposure.

Carpets made of wool are sold with "pestproof" guarantees which last some 10 years. In these carpets pesticides are applied during the manufacturing process to protect them from moths and keratin-digesting beetles. A concentration of permethrin of 60–180 mg kg^{-1} for impregnating woolen carpets is recommended [32].

High concentrations of permethrin in dust thus may either be due to a pest-control action performed by the occupants or commercial pest controllers or be due to the impregnation of carpets or textiles. The impregnation of woolen carpets with permethrin may cause concentrations in household dust of up to about 100 mg kg^{-1} [33]. As a consequence permethrin concentrations in house dust were significantly higher in homes with woolen carpets or woolen floor-to-floor coverings compared with those not having this household characteristic [34]. A similar situation is observed regarding DDT by comparing residues in homes with and without woolen carpets (Fig. 1).

Although PCP has been banned in Germany since 1989, and the last indoor application resulted from activities prior to 1978, this biocide is still evaporating in indoor environments. Concentrations of PCP in house dust from households which used wood preservatives in the past were significantly higher than those without a wood preservation history [35].

3
Factors Determining Indoor Concentrations

The variability of biocides in indoor air and house dust is high. This variability is caused, on the one hand, by the conditions under which the sample were taken but, on the other hand, there are spatial variations within homes and the concentrations of biocides in an indoor environment are influenced by cultural and climatic factors. Furthermore concentrations may vary with season and temperature and there are trends downwards or upwards with respect to discontinued use or an increase in application.

A high variation of concentrations of PCP in air and dust on consecutive measurements in the same buildings was reported by Schnelle-Kreis et al. [36].

Air and dust samples were taken from 75 rooms in 30 buildings where PCP-containing wood preservatives were probably used. Sampling was repeated four times within 18 months. The variability of the PCP concentrations in air and dust was in the same range as the measured values. Monitoring pesticides (aldrin, dieldrin, chlordane, chlorpyrifos and heptachlor) in indoor air in at least two locations revealed significant differences between levels in different areas of the home [37]. Furthermore a spatial distribution of allethrin in air was observed for different sampling sites in the same room [38]. By contrast, Wright et al. [15, 16] reported that no significant differences were found between rooms in different living areas of homes.

Biocides applied in an indoor environment differ significantly with climatic and cultural factors [10], as a variety of pesticides is used in different countries. In the USA substances for fighting insects, cockroaches and termites, for example, aldrin, chlordane, chlorpyrifos, diazinon, dieldrin, propoxur and heptachlor, are the pesticides most often found in indoor environments [26, 39–42]. Aldrin, chlordane, dieldrin and heptachlor, however, could never be detected in house dust from Germany [43, 44]. The same pesticides as in the USA are applied in Australia, where, in order, heptachlor, dieldrin, chlordane, aldrin and chlorpyrifos were the most frequently detected indoor pesticides [37]. For other regions, for example, developing countries, DDT and hexachlorocyclohexanes (HCHs) to fight mosquitoes (and to control malaria) are still marketed and applied indoors [45]. DDT is also found in indoor air and house dust of homes of Arizona near the Mexican border [46], although the use of DDT has been banned for many years in the USA primarily because of its environmental persistence. But Mexico began a 10-year phase out of DDT only in 1997. Great regional differences in DDT concentrations are also found regarding former West Germany and former East Germany. Although banned in former West Germany since 1972, it is still found in house dust with concentrations up to 40 mg kg^{-1} [35]. But former East Germany, where DDT was applied for wood protection (Hylotox) till 1989, concentrations in house dust tend to be much higher [47–49]; it amounted to up to several grams per kilogram in settled dust from attics [48]. In Germany public discussion nowadays mainly focuses on wood preservatives like PCP and lindane and insecticide pyrethroids, mainly permethrin.

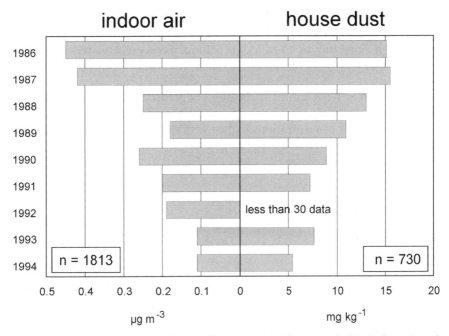

Fig. 2 Trend of concentrations of pentachlorophenol (90th percentiles) in indoor air and house dust from private houses in northern Germany where wood preservatives had been used (last application of pentachlorophenol in 1978)

Regarding fluctuations of indoor air concentrations of biocides with temperature the results are contradictory. Seasonal variations, i.e., a correlation with the average monthly temperature, were observed for indoor air chlordane concentrations [50]. They were high in summer and low in winter. A very strong correlation between average weekly air temperatures, measured over a range of –29.3 to +20.0 °C, and the logarithm of the average weekly concentration of PCP outdoors was also reported [51]. No correlation, however, between cypermethrin or chlorpyrifos levels and the temperature or relative humidity in the rooms was found by Wright et al. [15, 52].

Discontinued use or the replacement of one biocide by other compounds may result in downward or upward trends for biocides indoors. Fenske et al. [53], measuring chlorpyrifos and parathion in residences in a central Washington State agricultural community, reported the parathion concentrations in house dust to decrease tenfold from 1992 to 1995, consistent with the discontinued use of this product in the region in the early 1990s.

In Germany the usage of PCP for wood preservation indoors ceased in 1978. A decrease in the concentrations in indoor air and house dust was observed thereafter. The trend for the years 1986–1994 is displayed in Fig. 2.

Decreasing indoor concentrations for PCP were also visible regarding house dust from data of the German Environmental Survey (GerES) [54–56]. Concentrations of PCP in indoor air of rooms, where PCP was applied as a wood preservative, are nowadays normally of the same magnitude as in rooms without a PCP history. Indoor air of rooms without wood preservatives show concentrations of less than 0.025 µg m^{-3} [57] and less than 0.05 µg m^{-3} [58], respectively.

Opposite to PCP, the indoor applications of permethrin in Germany seem to be growing, as for many applications permethrin is replacing other insecticides. An increase in the concentrations in dust with time was observed as reported in the GerES in 1990–92 and 1998 [54, 59].

4
Analytical Methods

4.1
Air

General aspects of strategies to measure biocides in indoor air are given in guidelines, for example, by the VDI (Association of German Engineers) [60, 61].

Sampling of biocides like aldrin, allethrin, chlordane, chlorpyrifos, cyfluthrin, cypermethrin, p,p'-DDE, p,p'-DDT, deltamethrin, diazinon, dichlorvos, dieldrin, endrin, heptachlor, hepatchlorepoxide, lindane (γ-HCH), PCP, permethrin, o-phenylphenol, propoxur and tetramethrin in air is generally performed by passing air through an adsorbent which traps the chemicals. Different materials have been utilized to collect pesticides, for example, Chromosorb 102 [38, 62–66], Orbo 42 [52], Tenax [40, 67], Ambersorb XE-340 [68], silica gel [69–71], Florisil [47] and polyurethane foam [42, 49, 51, 72–76]. Particle-bound pesticides like 2,4-D or permethrin may be collected by glass, quartz or activated carbon-fiber filters [3, 42, 74]. Experiments by Roper and Wright [77], who generated vapors of 100 µg m^{-3} chlorpyrifos, chlordane, diazinon, propoxur and resmethrin in an air stream, showed no significant difference for the retention efficiency of five sorbents tested. Regarding the sampling efficiency, Chromosorb 102 and polyurethane foam appeared to be superior to other sorbents.

After air sampling the sorbents are subsequently extracted (seldom thermally desorbed) and the pesticides were analyzed using gas chromatrography (GC) or high-pressure liquid chromatography (HPLC). For HPLC quantification is done by UV absorption at an appropriate wavelength, whereas for GC either an electron capture detector (ECD) or a mass spectrometer (MS) are used for quantification. Depending on the sampled air volume detection limits for HPLC methods are about 0.1–1 µg m^{-3}, for GC–ECD methods about 0.01–0.02 µg m^{-3} and for MS methods about 0.001 µg m^{-3}.

For higher concentrations of pesticides in air impinging methods are an alternative to sorbent methods. PCP concentrations were analyzed after passing the air through an aqueous alkaline solution [36, 78–81]; DDT and lindane were passed through ethylene glycol [45].

It is important in active sampling to choose appropriate sorbents and methods. Lindane (γ-HCH), for example, is only present in the vapor phase, not in the particle phase [82]. Singh et al. [45] also reported lindane to be chiefly found in the vapor phase. DDT in contrast, when analyzed in rooms treated with DDT according to the National Malaria Eradication Program of India, was 34–78% particle-bound.

4.2
House Dust

Owing to its varying sources house dust is very heterogeneous and consists of a variety of inorganic and organic particles as well as fibers of different sizes [83–86]. As deposited dust (mainly floor dust) is most often collected by vacuuming, "house dust" is often regarded to be the content of vacuum cleaner bags.

The quantity and composition of house dust varies greatly with seasonal and environmental factors. According to Butte and Walker [58] the portion of organic matter in house dust samples varies between less than 5% and more than 95%; Fergusson et al. [87] reported the organic content of house dust between 25.7% and 56.5% and floor dust from Danish offices had a mean organic fraction of 33% [88].

House dust may adsorb pollutants released from activities and materials from inside the home or contaminants may be tracked in on particles and fibers from outside. Once adsorbed, contaminants do not degrade or degrade more slowly compared with outdoor environments [89]. Thus, house dust is a sink and repository; it collects pesticides like a passive sampler.

Mostly pesticides in house dust are analyzed from floor dust taken with vacuum cleaners (or other suction devices). Even though there are standard protocols for sampling house dust [84, 90, 91] a great variety in sampling techniques has to be noted. These were reviewed by Butte and Heinzow [85] and Macher [92].

After sampling the crude dust the vacuum cleaner bag has to be processed to obtain a subsample appropriate for analysis. Obtaining a more homogeneous subsample may be achieved by discarding larger particles like hairs and feathers, atypical objects, and coarse material (like clips, small toys, knobs and small stones) from the gross dust, or by choosing just "pieces of fluff" (dust bunnies) for analysis, as well as by sieving the dust [24, 26, 35, 54, 59, 93–98]. On sieving, coarse material is removed and a separation of particles from fibers is performed.

For an instrumental analysis of organic pesticides house dust is mostly extracted with organic solvents [93, 99, 100] and the dissolved organic components are separated and quantified by chromatographic methods [35, 93, 99–102] either with or without sample cleanup [103]. In the case of the analysis of polar pesticides such as phenoxy acid herbicides or PCP [35, 99–101] derivatization might be recommended.

Aside from instrumental methods, enzyme-linked immunosorbent assay for PCP and 2,4-D is commercially available but positive biases for 2,4-D as well as for PCP were seen in most house dust samples [104]. Thus, the enzyme-linked immunosorbent assay cannot be recommended as a quantitative analytical tool, but might be useful as a screening tool [105].

Detection limits for many pesticides in house dust form vacuum bags have been reported to be about 0.1 mg kg^{-1} [35, 96].

5
Occurrence of Pesticides Indoors

5.1
Air

High concentrations of pesticides in air were measured mainly during the application, especially on spraying or nebulizing pesticide formulations, during electroevaporation or on using insecticide strips. For pyrethroids concentrations may reach some hundred micrograms per cubic meter during spray operations [70] and on electroevaporation some micrograms per cubic meter [70]. For organophosphates such as dichlorvos even concentrations in the milligrams per cubic meter range may be reached as shown by Sagner and Schöndube [106] using experimental chambers resembling living rooms and by Weis et al. [65], who put down insecticide strips containing dichlorvos in an experimental room. During spraying persons may suffer absorption by inhalation and percutaneously. The respiratory exposure of applicators to airborne concentrations of chlorpyrifos used to fight termites averaged 8.1% of the threshold limit value (chlorpyrifos threshold limit value is 200 µg m^{-3}) [14].

Some time after spraying insecticides or painting wood, high concentrations in air are possible. Warren et al. [69] obtained concentrations for PCP in air vaporized from wood up to 580 µg m^{-3}. These values are of the same magnitude as concentrations (up to 160 µg m^{-3}) reported by Gebefügi et al. [67] for PCP in indoor air some days after painting timber with wood preservatives. The amount of pesticide vaporized from wood into the surrounding air, i.e., the quantity released per unit time, is dependent on the formulation of the wood preservative and the timber species [107]. The concentration of PCP in indoor air is further controlled by the quantity of wood present in the room in relation to its volume, by the exchange rate of the indoor air and the temperature in the room.

After application a decrease of the concentration in air with time is usually observed. Wright et al. [15–17] published chlorpyrifos concentrations in indoor air 8 years after termite control. Concentrations starting with 8.5 µg m^{-3} after the application [15] were up to 9 µg m^{-3} 4 years later [16] and than dropped to 0.7 µg m^{-3} 8 years thereafter [17]. The concentration of cypermethrin decreased within 84 days from 18.2 to 0.3 µg m^{-3} in the air of vacant dormitory rooms following its application for cockroach control [52]. The half-life of allethrin in indoor air was reported to be only 1.9 days [38]. The use of pest strips to protect rooms from insects may lead to chronic exposure for several weeks. After application of an insect trip containing 16 g dichlorvos in an experimental test chamber (34 m^3) the dichlorvos concentration reached a maximum after 7 days with a concentration of 1,292 µg m^{-3} and dropped under 1 µg m^{-3} only after 95 days [65].

Typical concentrations for pesticides in indoor air obtained under defined conditions either directly during or after application and up to 20 years after an incident were reported by Pluschke [108]. Concentrations of various pesticides in the air of nonoccupational indoor environments of different countries and measured under different conditions are compiled in Table 2. Additionally results for chlorpyrifos in air were reviewed by Lemus and Abdelghani [109].

Regarding the ubiquitous contamination of indoor, air heptachlor was the pesticide with the highest concentrations of 13 target pesticides (including organophosphate and organochlorine pesticides as well as the phenoxy herbicide 2,4-D). Samples were taken in rooms of childcare centers in central North Carolina in 1997 [42]. Out of 33 target pesticides mean concentrations of dichlorvos, heptachlor, chlorpyrifos, chlordane, propoxur and diazinon in indoor air of households in Jacksonville, Florida, exceeded 100 ng m^{-3} [26]. Samples were taken in households in which no household member was employed in a position in which the primary activity involved the handling or use of pesticides [26]. For a compilation of concentrations of pesticides in indoor air obtained under different circumstances see Table 2.

Table 2 Concentrations of pesticides in the air of nonoccupational indoor environments

Compound	Maximum concentration	Circumstances	Reference
Aldrin	0.223 ng m^{-3}	Childcare centers (chosen to assess the ubiquitous occurrence of chemicals), USA	Wilson et al. (2001) [42]
Allethrin	5 µg m^{-3}	12 h after using an electroevaporator	Claas and Kintrup (1991) [70]
Allethrin	48 ng m^{-3}	0.1 days after application (distribution by vending machines)	Eitzer (1991) [38]
Chlordane	0.81 µg m^{-3}	Living areas of homes, 1 year after termiticide application	Louis and Kisselbach (1987) [157]

Table 2 (continued)

Compound	Maximum concentration	Circumstances	Reference
α-Chlordane	30.7 ng m^{-3}	randomly selected homes, USA	Roinestadt et al. (1993) [40]
α-Chlordane	19.3 ng m^{-3}	→ aldrin	Wilson et al. (2001) [42]
γ-Chlordane	44.7 ng m^{-3}	→ α-chlordane	Roinestadt et al. (1993) [40]
γ-Chlordane	29.1 ng m^{-3}	→ aldrin	Wilson et al. (2001) [42]
Chloronaph-thalenes	108 µg m^{-3}	School rooms with wooden parts treated with preservatives	Pluschke et al. (1996) [64]
Chlorpyrifos	150 ng m^{-3}	→ α-chlordane	Roinestadt et al. (1993) [40]
Chlorpyrifos	0.7 µg m^{-3}	8 years after termite control, USA	Wright et al. (1994) [17]
Chlorpyrifos	21.7 ng m^{-3}	→ aldrin	Wilson et al. (2001) [42]
Cyfluthrin	90 µg m^{-3}	15 s after spraying	Claas and Kintrup (1991) [70]
Cyfluthrin	<5 ng m^{-3}	After fighting cockroaches (up to four times per year in up to 5 years), Germany	Ball et al. (1993) [74]
Cypermethrin	5 ng m^{-3}	→ cyfluthrin	Ball et al. (1993) [74]
Cypermethrin	19 µg m^{-3}	Immediately after spraying, USA	Wright et al. (1993) [52]
Chlorothalonil	0.47 ng m^{-3}	→ aldrin	Wilson et al. (2001) [42]
p,p'-DDE	0.591 ng m^{-3}	→ aldrin	Wilson et al. (2001) [42]
p,p'-DDT	14.6 µg m^{-3}	Rooms treated with either DDT or lindane for mosquito control, India	Singh et al. (1992) [45]
p,p'-DDT	5 µg m^{-3}	After using DDT for wood preservation, Germany	Lederer and Angerer (1997) [158]
p,p'-DDT	4.6 µg m^{-3}	Attics, more than 20 years after application of wood preservatives, German	Rosskamp et al. (1999) [49]

Table 2 (continued)

Compound	Maximum concentration	Circumstances	Reference
p,p'-DDT	0.882 ng m^{-3}	→ aldrin	Wilson et al. (2001) [42]
Deltamethrin	<10 ng m^{-3}	→ cyfluthrin	Ball et al. (1993) [74]
Diazinon	6 ng m^{-3}	→ α-chlordane	Roinestadt et al. (1993) [40]
Diazinon	62.4 ng m^{-3}	→ chlorpyrifos	Wilson et al. (2001) [42]
Dichlorvos	250 ng m^{-3}	→ α-chlordane	Roinestadt et al. (1993) [40]
Dichlorvos	1,292 µg m^{-3}	Experimental room (34 m^3), 1 week after using an insecticide strip	Weis et al. (1998) [65]
Dieldrin	1.56 ng m^{-3}	→ aldrin	Wilson et al. (2001) [42]
Endrin	3.26 ng m^{-3}	→ aldrin	Wilson et al. (2001) [42]
Heptachlor	0.57 µg m^{-3}	→ chlordane	Louis and Kisselbach (1987) [157]
Heptachlor	336 ng/m^3	→ aldrin	Wilson et al. (2001) [42]
Lindane	2 µg m^{-3}	Rooms where wood y preservatives were used (at least 7 years after application), German	Blessing and Derra (1992) [71]
Lindane	2,000 µg m^{-3}	→ p,p'-DDT, India	Singh et al. (1992) [45]
Lindane	0.93 µg m^{-3}	→ p,p'-DDT	Rosskamp et al. (1999) [49]
Lindane	13.1 ng m^{-3}	→ aldrin	Wilson et al. (2001) [42]
Pentachloro-phenol	160 µg m^{-3}	Model room (swimming hall with wood, treated with pentachlorophenol, Germany)	Gebefügi et al. (1979) [67]
Pentachloro-phenol	82 µg m^{-3}	1.5 years after using wood preservatives, Germany	Dahms and Metzner (1979)
Pentachloro-phenol	25 µg m^{-3}	<9 years after using wood preservatives, Germany	Krause and Englert (1980) [136]

Table 2 (continued)

Compound	Maximum concentration	Circumstances	Reference
Pentachloro-phenol	2 µg m^{-3}	→ lindane	Blessing and Derra (1992) [71]
Pentachloro-phenol	576 ng m^{-3}	Buildings assumed to have used wood preservatives (more than 20 years ago), Germany	Schnelle-Kreis et al. 2000 [81]
Pentachloro-phenol	17.6 ng m^{-3}	→ aldrin	Wilson et al. (2001) [42]
Permethrin	8.8 µg m^{-3}	14 months after fighting cockroaches, Germany	Fromme (1991) [159]
Permethrin	4.7 ng m^{-3}	→ cyfluthrin	Ball et al. (1993) [74]
Permethrin	322 ng m^{-3}	Rooms showing high permethrin concentrations in house dust, Germany	Stolz et al. (1994) [160]
Permethrin	85.2 ng m^{-3}	Private homes with woolen floor-to-floor covering, Germany	Berger-Preiss et al. (2002) [33]
o-Phenyl-phenol	58 ng m^{-3}	→ α-chlordane	Roinestadt et al. (1993) [40]
Propoxur	63 ng m^{-3}	→ α-chlordane	Roinestadt et al. (1993) [40]
Tetramethrin	300 µg m^{-3}	15 s after spraying	Claas and Kintrup (1991) [70]

5.2
House Dust

House dust is a sink and reservoir for semivolatile and particle-bound pesticides. Thus, it is a measure for the average contamination of an indoor environment. Pesticides found in house dust are those that are (1) stable in the indoor environment and (2) regularly applied in formulations to fight pests indoors. In contrast to an outdoor environment, where modern pesticides are degraded rather quickly by microorganisms, hydrolysis and UV light, pesticides used indoors tend to be persistent [85]. Thus, they are a reservoir for chronic exposure.

As already mentioned a high variability in pesticide residues in house dust is observed as different sampling and sample preparation techniques result in divergent concentrations. In general, concentrations increase with decreasing particle size [58, 110, 111].

Table 3 Pesticides in German house dust (mg kg^{-1})

Compound	Median	95th percentile	Reference
Chlorpyrifos	≤0.1	0.63	Walker et al. (1999) [35][a]
Chlorpyrifos	≤0.05	0.70	Becker et al. (2002) [54][b]
p,p'-DDT	0.31	4.2	Walker et al. (1999) [35][a]
p,p'-DDT	<0.05	1.2	Becker et al. (2002) [54][b]
Lindane	≤0.1	0.83	Walker et al. (1999) [35][a]
Lindane	≤0.05	0.75	Becker et al. (2002) [54][b]
Methoxychlor	0.92	27	Walker et al. (1999) [35][a]
Methoxychlor	≤0.05	5.8	Becker et al. (2002) [54][b]
Pentachlorophenol	0.95	8.0	Walker et al. (1999) [35][a]
Pentachlorophenol	0.2	2.9	Becker et al. (2002) [54][b]
Permethrin (Σ cis+trans)	0.67	37	Walker et al. (1999) [35][a]
Permethrin (Σ cis+trans)	0.17	14.5	Becker et al. (2002) [54][b]
Piperonylbutoxide	0.11	13	Walker et al. (1999) [35][a]
Piperonylbutoxide	0.04	3.7	Becker et al. (2002) [54][b]
Propoxur	≤0.1	0.90	Walker et al. (1999) [35][a]
Propoxur	≤0.1	0.6	Becker et al. (2002) [54][b]
Tributyltin	0.5	2.8	Haumann and Thumalla (2002) [161][c]

[a] Homes from North Rhineland Westphalia and Lower Saxony; house dust collected by the householders with their own vacuum cleaners sieved to 63 µm or smaller (n=336).
[b] Homes from eastern and western Germany; house dust collected by the householders with their own vacuum cleaners sieved to 2 mm or smaller (n~740).
[c] Dust samples exactly 7-days old, collected by vacuuming according to VDI guideline 4300, part 8 [84], no sieving (n=33).

Pesticides found in house dust may be chemically classified as organo-chlorine compounds, like chlordane, DDT, dieldrin, lindane, heptachlor and methoxychlor [24, 26, 35, 40–44, 46–49, 54, 56, 58, 96, 112–114], organo-phosphorous pesticides, like chlorpyrifos, diazinon, dichlorvos, isofenfos and malathion [5, 24, 26, 35, 40, 42, 44, 54, 112, 115, 116], pyrethroids, like cyfluthrin, cypermethrin and permethrin [35, 40, 44, 54, 56, 59, 95, 110, 112, 114, 117, 118], carbamates, like bendiocarb, carbaryl and propoxur [35, 40, 44, 54, 112], and phenols, like PCP [24, 35, 43, 44, 54, 56, 58, 94, 96, 114, 119], chlorocresol [120], and *o*-phenylphenol [26, 40, 112].

Residues of pesticides from representative collectives concerning Germany and the USA are compiled in Tables 3 and 4. Ninety fifth percentiles represent a measure for the ubiquitous indoor contamination in these countries.

Pesticide concentration in house dust samples obtained because of certain incidents (fighting cockroaches, wood preservation, etc.) were reviewed by Butte [111]; results of pesticide analyses in house dust were also given by Lioy et al. [5] and Butte and Heinzow [85]. For further details on pollutants including pesticides in indoor environments see Chap 5.

Table 4 Pesticides inhouse dust from the USA (mg kg^{-1})

Compound	Median	95th percentile
Carbaryl	0.050	3.01
α-Chlordane	≤0.021	0.44
γ-Chlordane	0.021	0.69
Chlorpyrifos	0.108	3.53
4,4'-DDE	0.023	0.20
4,4'-DDT	0.090	1.6
Diazinon	0.025	0.76
Methoxychlor	0.074	2.07
cis-Permethrin	0.33	20.9
trans-Permethrin	0.70	38.7
o-Phenylphenol	0.25	1.25
Propoxur	0.072	1.45

Homes from the Detroit area, Michigan, the entire state of Iowa, Los Angeles County, and the Seattle area, Washington; house dust from bags in current use sieved to 150 μm or smaller (n~600). Camann et al. (2002) [112].

5.3
Associations of Pesticides in Indoor Air and House Dust

Roinestadt et al. [40], analyzing 23 pesticides in indoor air and dust, reported that pesticides in air were always found in the corresponding dust with the exception of dichlorvos, o-phenylphenol and chlordane. The majority of household pesticides, however, are preferably detected in the home environment by dust sampling [40]. This holds true particularly for permethrin, which could not be detected in the air (detection limit 1 ng m^{-3}), whereas it was present in the dust samples in the milligrams per kilogram range [40]. Stolz et al. [110] reporting results for permethrin in dust samples and air observed no correlation.

Similar results were reported by Berger-Preiss et al. [33], who analyzed permethrin in the indoor air of 80 private homes in Hannover (Germany). A maximum of 15.2 ng m^{-3} was observed, but the concentration of permethrin in air was highly dependent on the concentration of suspended particles. Berger-Preiss et al. [33] concluded that the permethrin concentration in air was mainly influenced by the degree of carpet fiber abrasion, as their investigations were performed in rooms covered with woolen carpets impregnated with permethrin.

For the wood preservative PCP, however, there was a relation between concentrations in air and in dust during the first 2 years after painting timber [121]. After this time this relation was no longer observed. No correlation for PCP in air and dust measured at least 20 years after the application was reported by Liebl et al. [119], but Schnelle-Kreis et al. [81] still found a highly significant correlation of PCP in freshly settled house dust and indoor air.

The data suggest that semivolatile pesticides (e.g., chloronaphthalene, dichlorvos, lindane and pentachorophenol) might be present in the air as well as in dust. For nonvolatile, i.e., particle-bound, pesticides, however, house dust is the material of choice to indicate indoor contamination. As house dust is a "long-term accumulative sample" trapping, accumulating and preserving contaminants [113], and has been regarded as an "indoor-pollution archive" [58], it is especially useful to detect pesticides applied in the past.

6
Impacts of Pesticides in Indoor Environments

6.1
Indoor Pesticides and Health

Indoor contamination is one source of exposure to toxic pollutants and has been classified as a high environmental risk [6, 122]. Regarding the principal exposure routes, inhalation, dietary ingestion, dermal and nondietary ingestion, pesticides in indoor air and bound to suspended particles may contribute to exposure via inhalation, house dust to dermal penetration and nondietary ingestion, respectively. Health effects have been closely associated with indoor pesticide applications and pesticides in house dust. Small children are considered to be the population at highest risk since they spend most of their time indoors and much of this time is spent in contact with floors, engaging in mouthing of hands, toys and other objects [18]. Recent findings of indoor exposure indicate that young children are at higher risks to semivolatile pesticides than had been previously estimated [123], as they live closer to the ground [124]. Health effects linking indoor exposure to pesticides include childhood leukemia [125, 126–130], developmental inhibition [6], reduction in motor skills, coordination and attention disorders [131] as well as neuroblastoma [132]. But improvements in measures of household exposures relating to health effects are still needed [133].

6.2
Exposure Assessment for Pesticides in Indoor Air and House Dust

6.2.1
Indoor Pesticides and Human Biomonitoring

The input of pesticides in an indoor environment may result from either a direct application (fighting, e.g., insects like flies, mosquitoes, and fleas) or pesticides used in the preservation of wood and timber, pesticides of textile finishing, and finishing of leather, carpets, fabric, etc., pesticides (mainly fungicides) in varnishes, colors, adhesives, or pesticides brought in by foot traffic, through pets, etc. from outdoors. Pesticides then spread in indoor air and in house dust and are adsorbed to surfaces, i.e., pesticide residues in a home environment are likely to contribute to human exposure.

Recent findings of indoor exposure studies of chlorpyrifos indicate that young children are at high risk to this semivolatile pesticide [134]. Even after a single broadcast use of chlorpyrifos by certified applicators in apartment rooms, it continued to accumulate on children's toys and hard surfaces 2 weeks after spraying. The estimated chlorpyrifos exposure levels from indoor spraying for children were estimated to be approximately 21–119 times above the current recommended reference dose from all sources [123].

Levels of the chlorpyrifos urine metabolite were found to be significantly correlated with chlorpyrifos concentrations in air and dust [135].

House dust serves as a reservoir for pesticides in households [85]. Dust ingestion scenarios show that exposures could also exceed the diazinon chronic reference dose [115]. Support for the thesis that household dust may not only be a direct exposure path but may serve as an indicator for all indoor exposure paths can be concluded from correlations between pesticides in dust and in samples of human origin. Regarding PCP, a semivolatile pesticide, concentrations in urine of women and children corresponded well with indoor dust samples from vacuum cleaner bags [13, 136].

A significant correlation between PCP concentrations in passively deposited particulate matter and in urine was further reported by Meissner and Schweinsberg [137]. On the other hand, no correlation was observed for PCP in household dust and blood by Liebl et al. [119], or for PCP in dust and urine by Rehwagen et al. [94]. Further, no association between concentrations in house dust or inhalable suspended particles in indoor air and of metabolites in urine was found for permethrin, a nonvolatile (particle-bound) pesticide [33, 138].

Contradictory results for associations between pesticide levels indoors (air, dust) and results form human biomonitoring may be due to different volatilities of the pesticides and may be determined by the magnitude in contamination levels. For semivolatile pesticides it may be easier to detect an association, as indoor air and house dust may serve for exposure in contrast to particle-bound pesticides with house dust as the only exposure path. Furthermore high contamination levels make it easier to detect an association, as with low indoor contamination levels associations may be hidden by the ubiquitous presence of pesticides in indoor environments and by nonindoor exposure pathways like dietary intake.

Besides the magnitude of pesticide residues indoor exposure might be influenced by household characteristics. Residents with large carpeted areas within their dwellings had a higher exposure to diazinon and chlorpyrifos for all routes versus those in less carpet-covered areas [139].

Children of agricultural families had a higher potential for exposure to organophosphorous pesticides (azinphosmethyl, chlorpyrifos, parathion and phosmet) than children of nonfarm families [30]. Dialkyl phosphate metabolites of organophosphorous pesticides measured in children's urine were also elevated for agrarian children when compared with reference children [53]. Tracking-in might be the main factor, as soil and house dust concentrations of these pesticides were elevated in homes of agrarian families (household members engaged in agricultural production) when compared with nonagricultural reference homes in the same community [53].

According to Bradman et al. [115], additional research is feasible and needed to assess the magnitude and distribution of these risks from indoor exposure to pesticides. An excellent review on dust as a metric for use in residential and building exposure assessment and source characterization was recently published by Lioy et al. [5].

6.2.2
Exposure Pathways

Gaseous pesticides are evenly dispersed in the air. In the case of inhalation, the anatomy and physiology of the respiratory system diminishes the pesticide concentration in inspired air. As pesticides are mostly lipid-soluble, they are usually not removed in the upper airways but tend to deposit in the distal portion of the lung, the alveoli [83] and may then be absorbed into the blood stream.

Particulate pesticides or particle-bound pesticides are either dispersed in the air or they are deposited as house dust. They may enter the human body either by inhalation or through oral intake, i.e., nondietary ingestion of dust (infants, toddlers) and ingestion of particles adhering to food, to surfaces in homes (e.g., toys) and to the skin. Pesticides adsorbed to surfaces may further be absorbed directly through the skin by dermal contact.

Dust suspended in air and thereby inhaled is deposited in different parts of the alveolar tract (nose, throat and lung) dependent upon its size (aerodynamic diameter). The efficacy of deposition in the alveolar tract increases with decreasing particle size. Particles smaller than 10 µm enter the tracheo-bronchial area, reaching, depending on their size, the trachea, bronchi or alveoli. Substances adsorbed to dust particles that enter the alveoli can be absorbed by epithelial cells of the lung or through macrophagial phagocytosis.

But exposure to house dust does not exclusively occur and may not even occur predominantly via inhalation. For instance, ingestion of house dust particles adhering to food objects and the skin or direct absorption through the skin may be primary routes of exposure [24]. This holds true especially for small children as they have the tendency to be very tactile and handle and place nonfood objects in their mouths [140]. Regarding mouthing behavior, the daily frequency of both mouth and tongue contacts with hands, other body parts, surfaces, natural objects and toys was evaluated by Tulve et al. [141] for children from 11 to 60 months of age. A clear relationship was observed between mouthing frequency and age. Children less than 24-months old exhibited a frequency of mouthing with an average of 81 events per hour and children more than 24 months of age 42 events per hour.

Estimations of the quantity of dust ingested daily vary. Infants and toddlers, who crawl and mouth their hands and other objects, ingest a daily rate twice that of adults and the amount ingested is estimated to be 0.02–0.2 g [142]. Toddlers, possibly eating nonfood items, may consume as much as 10 g of soil or dust per day ("pica-behavior") [142]. US-EPA assumptions on the indoor exposure among young children (i.e., 2.5-years old) were based on an ingestion of 100 mg of house dust per day during the winter months and of 50 mg during the warmer months [83]. In Germany the amount of dust ingested is estimated to be 20–100 mg day^{-1} for 1–6-year-old children, 5–25 mg day^{-1} for 7–14-year-old children and 2–10 mg day^{-1} for 15–75-year-old adults [143]. The daily amount of carpet dust ingested by children was assumed to be 100 µg day^{-1} [144].

The quantity of suspended dust inhaled and the intake of pesticides adsorbed to it may be calculated rather exactly, and the intake of pesticides from deposited dust via oral pathways may at least be roughly estimated. But data supporting the amount of chemicals absorbed through dermal contact of contaminated house dust and through direct contact to contaminated surfaces are still lacking. Semivolatile pesticides like chlorpyrifos will accumulate not only in house dust but also on toys and on sorband surfaces [134]. Data from the NHEXAS of Arizona support the importance of dermal penetration of semivolatile pesticides like chlorpyrifos and diazinon as a route of residential human exposure [116].

6.3
Reference and Guideline Values to Assess Indoor Quality

For the interpretation of results obtained from monitoring pesticides in indoor environments in general two approaches might be used: (1) the comparison with reference values and (2) the application of a risk-assessment methodology related to the hazard of the compound, leading to standards or guideline values. The former approach will help determine whether a measured value is "normal", but has no health meaning or regulatory implication.

Reference values are intended to characterize the upper margin of the current background contamination of a pollutant at a given time. Regarding pesticides in indoor environments, reference values may be calculated by applying the procedure as for environmental toxins in body fluids [145]. This concept defines the upper margin (95th percentile) of the concentration levels as reference value. To be valid for the general population, reference values must be derived from studies large enough to be representative of that population. Reference values may show trends, as contaminant levels may change with time. The extent, distribution and determinants of pesticides in house dust have been evaluated by the GerES [56, 59, 146] or the NHEXAS [139, 147–149] and the Minnesota Children's Pesticide Exposure Study [150, 151] in the USA. Data collected for pesticides in house dust from national surveys and from studies, which are regarded to be representative for the study population are compiled in Tables 3 and 4.

It must be emphasized that reference values have no uniform character and relate to the reference population only. They are subject to regional/spatial and temporal variation and, in the case of house dust, are significantly governed by sampling technique and sample preparation (e.g., sieving).

Reference values for pesticides in indoor air are not available. Studies obtained under specific criteria for the selection of these samples are given in Table 2. Some of them may give an impression of the ubiquitous occurrence of pesticides in indoor air.

Applying a risk-assessment methodology related to the hazard of a pesticide leads to standards or guidelines. Regarding levels or concentrations of pesticides in a private indoor environment, there are no standards, i.e., values which might be legally enforced. Guideline values, however, have been defined for PCP in indoor air [152] and tentative benchmarks have been suggested for pesticides in house dust [85].

In Germany a general scheme for deducing guideline values for indoor air (Richtwerte für Innenraumluft) was developed by the Indoor Air Commission of the German Federal Environmental Agency [153, 154]. It is based on toxicity data of the compound, if possible the lowest observable adverse effect level; for the general scheme of defining guideline values see Scheme 1.

The concept of guideline values for indoor air involves two levels. Guideline value II is a health-related value based on current toxicological and epidemiological knowledge. If the concentration of a pollutant in indoor air reaches or exceeds this value immediate action has to be taken to avoid health impairment. For concentrations of pollutants falling below guideline level I, they do not give rise to adverse health effects even under lifelong exposure. It may further be regarded as the level to be reached after the sanitization of rooms (diminishing the indoor contamination after exceeding guideline value II). For further details on guideline values for indoor air see Refs. [153, 154]; a review on guideline values for indoor air reflecting various countries was given by Pluschke [155].

Risks associated with the ingestion of contaminated dust have been estimated by Butte and Heinzow [85] using the chronic oral reference dose available from the US-EPA Integrated Risk Assessment Information Service [156]. With a focus on small children (age 1–6 years, mean body weight 16 kg) and a daily intake of 100 mg house dust [24, 83] tentative benchmarks for house dust were calculated. The assessment indicated for chlorpyrifos, DDT and diazinon that the tolerable exposure concentration in house dust might be exceeded in some samples and chlorpyrifos especially can be considered a potential hazard to householders.

Guideline Values for Indoor Air Pollutants
(Application to Pentachlorophenol)

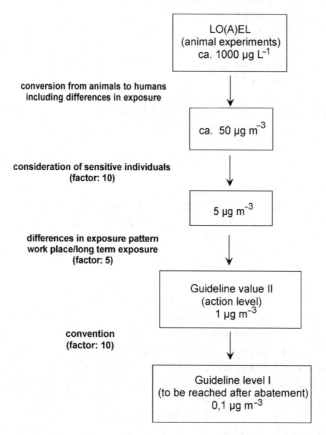

Scheme 1 Derivation of guideline values for indoor air (example pentachlorophenol)

7
Summary

In this review the use of indoor air and of house dust for identifying indoor contaminants and for characterizing the potential exposure were discussed. Furthermore a short overview of methods to analyze pesticides in air and dust as well as results was given. Results for indoor concentrations of pesticides may vary significantly with the conditions under which they are obtained.

Pesticides are either SVOCs or POM. If they belong to SVOCs, an association of concentrations in indoor air and house dust might be observed. To get an impression of indoor contamination consecutive measurements in indoor air or in house dust, that serves as a reservoir for pesticides, should be performed. Indoor air is the ideal material to get information about the actual indoor contamination and the exposure via inhalation. House dust is a measure of chronic exposure. It may be regarded as a pollution archive as it is traps, accumulates and preserves pesticides tracked in, applied or brought in by treated items. Ingestion and dermal contact with house dust can be the primary routes of chronic exposure to pesticides.

Impacts of pesticides in indoor air and house dust were discussed with respect to reference values and guideline values. In view of this discussion it has to be emphasized that an application of pesticides in indoor environments has to be well-considered and, if possible, either avoided or minimized to the greatest possible extent.

References

1. Environment and Health Research for Europe (1999) Background document to the 3rd ministerial conference of environment and health (London, 16–18 June 1999). WHO Regional Office for Europe, Copenhagen
2. Schulz C, Becker K, Friedrich C, Helm D, Hoffmann K, Krause C, Seifert B (1999) Epidemiol 10:200P
3. Klepeis NE, Nelson WC, Ott WR, Robinson JP, Tsang AM, Switzer P, Behar JV, Hern SC, Engelmann WH (2001) J Exposure Anal Environ Epidemiol 11:231
4. Jones AP (1999) Atmos Environ 33:4535
5. Lioy PJ, Freeman NC, Millette JR (2002) Environ Health Perspect 110:969
6. Roberts JW, Dickey P (1995) Rev Environ Contam Toxicol 143:59
7. World Health Organisation (1989) Indoor air quality: organic pollutants. Euro report and studies no 111. WHO Regional Office for Europe, Copenhagen
8. Rippen G (2002) Handbuch Umweltchemikalien. Ecomed, Landsberg
9. Tomlin C (1994) The pesticide manual, 10th edn. The British Crop Protection Council, Surrey, and The Royal Society of Chemistry, Cambridge
10. Davis JR, Brownson RC, Garcia R (1992) Arch Environ Contam Toxicol 22:260
11. Adgate JL, Kukowski A, Stroebel C, Shubat PJ, Morrell S, Quackenboss JJ, Whitmore RW, Sexton K (2000) J Exposure Anal Environ Epidemiol 10:159
12. Hoppin J A, Yucel F, Dosemeci M, Sander, PD (2002) J Exposure Anal Environ Epidemiol 12:313
13. Krause CM, Chutsch M, Englert N (1989) Environ Int 15:443
14. Leidy RB, Wright CG, Dupree HE Jr (1991) Bull Environ Contam Toxicol 47:177
15. Wright CG, Leidy RB, Dupree HE Jr (1988) Bull Environ Contam Toxicol 40:561
16. Wright CG, Leidy RB, Dupree HE Jr (1991) Bull Environ Contam Toxicol 46:686
17. Wright CG, Leidy RB, Dupree HE Jr (1994) Bull Environ Contam Toxicol 52:131
18. Lewis RG, Nishioka MG (1999) Proceedings of indoor air '99, vol 2, p 416
19. Nishioka MG, Lewis RG, Brinkman MC, Burkholder HM (2002) Bull Environ Contam Toxicol 68:64
20. Nishioka MG, Burkholder HM, Brinkman MC, Lewis RG (1999) Environ Sci Technol 33:1359
21. Nishioka MG, Burkholder HM, Brinkman MC, Gordon SM (1996) Environ Sci Technol 30:3313
22. Morgan MK, Stout DM, Wilson NK (2001) Bull Environ Contam Toxicol 66:295
23. Nishioka MG, Lewis RG, Brinkman MC, Burkholder HM, Hines CE, Menkedick JR (2001) Environ Health Perspect 109:1185

24. Lewis RG, Fortmann RC, Camann DE (1994) Arch Environ Contam Toxicol 26:37
25. Roberts JW, Clifford WS, Glass G, Hummer PC (1999) Arch Environ Contam Toxicol 36:477
26. Whitmore RW, Immerman FW, Camann DE, Bond AE, Lewis RG, Schaum JL (1994) Arch Environ Contam Toxicol 26:47
27. Thatcher TL, Layton DW (1995) Atmos Environ 29:1487
28. Aurand K, Drews M, Seifert B (1983) Environ Technol Lett 4:433
29. Chiaradia M, Gulson BL, MacDonald K (1997) Occup Environ Med 54:117
30. Simcox NJ, Fenske RA, Wolz SA, Lee IC, Kalman DA (1995) Environ Health Perspect 103:1126
31. Fenske RA, Lu CS, Simcox NJ, Loewenherz C, Touchstone J, Moate TF, Allen EH, Kissel JC (2000) J Exposure Anal Environ Epidemiol 10:662
32. Klingenberger H (1994) VDI-Berichte 1122:387
33. Berger-Preiss E, Levsen K, Leng G, Idel H, Sugiri D, Ranft U (2002) Int J Hyg Environ Health 205:459
34. Butte W (2003) In: Morawska L, Salthammer T (eds) Indoor environment. Airborne particles and settled dust. Wiley VCH, Weinheim, p 407
35. Walker G, Hostrup O, Butte W (1999) Gefahrst Reinh Luft 59:33
36. Schnelle-Kreis J, Scherb H, Gebefugi I, Kettrup A (1999) J Environ Monit 1:353
37. Dingle P, Williams D, Runciman N, Tapsell P (1999) Bull Environ Contam Toxicol 62:309
38. Eitzer BD (1991) Bull Environ Contam Toxicol 47:406
39. Lewis RG, Bond AE, Johnson DE, Hsu JP (1988) Environ Monit Assess 10:59
40. Roinestad KS, Louis JB, Rosen JD (1993) J Assoc Off Anal Chem Int 76:1121
41. Wright CG, Leidy RB, Dupree HE (1996) Bull Environ Contam Toxicol 56:21
42. Wilson NK, Chuang JC, Lyu C (2001) J Exposure Anal Environ Epidemiol 11:449
43. Mattulat A (2002) VDI-Berichte 1656:551
44. Pöhner A, Simrock S, Thumulla J, Weber S, Wirkner T (1998) Z Umweltmed 6:337
45. Singh PP, Udeaan AS, Battu S (1992) Sci Total Environ 116:83
46. Robertson G, Lebowitz M, Needham L, O'Rourke MK, Rogan S, Petty J, Huckins J (2002) Proceeding of indoor air 2002, vol 4, p 63
47. Baudisch C, Prösch J (2000) Umweltmed Forsch Prax 5:161
48. Horn W, Rosskamp E, Ullrich D, Seifert B (1999) Proceedings of indoor air '99, vol 2, p 794
49. Rosskamp E, Horn W, Ullrich D, Seifert B (1999) Umweltmed Forsch Prax 4:354
50. Asakawa F, Jitsunari F, Takeda N, Manabe Y, Suna S, Fukunaga I (1994) Bull Environ Contam Toxicol 52:546
51. Waite DT, Gurprasad NP, Cessna AJ, Quiring DV (1998) Chemosphere 37:2251
52. Wright CG, Leidy RB, Dupree HE Jr (1993) Bull Environ Contam Toxicol 51:356
53. Fenske RA, Lu C, Barr D, Needham L (2002) Environ Health Perspect 110:549
54. Becker K, Seiwert M, Kaus S, Krause C, Schulz C, Seifert B (2002) Proceedings of indoor air 2002, vol 4, p 883
55. Krause C, Chutsch M, Henke M, Kliem C, Leiske M, Schulz C, Schwarz E (1991) WaBoLu-Hefte 2/1991. Institut für Wasser-, Boden- und Lufthygiene des Bundesgesundheits-amtes, Berlin
56. Seifert B, Becker K, Helm D, Krause C, Schulz C, Seiwert M (2000) J Exposure Anal Environ Epidemiol 10:552
57. Eckrich W (1989) VDI-Berichte 745:297
58. Butte W, Walker G (1994) VDI-Berichte 1122:535
59. Friedrich C, Becker K, Hoffmann G, Hoffmann K, Krause C, Nöllke P, Schluz C, Schwabe R, Seiwert M (1998) Gesundheitswesen 60:95

60. Verein Deutscher Ingenieure (1995) Indoor air-pollution measurement. General aspects of measurement strategy. VDI 4300, part 1
61. Verein Deutscher Ingenieure (1997) Indoor air-pollution measurement. Measurement strategy for pentachlorophenol (PCP) and γ-hexachlorocyclohexane (lindane) in indoor air. VDI 4300, part 4
62. Butte W (1987) Fresenius Z Anal Chem 327:33
63. Kasel U, Wichmann G, Juhl B (1995) Staub Reinh Luft 55:439
64. Pluschke P, Balzer W, Nix G, Schelle G (1996) Proceedings of indoor air '96, vol 3, p 71
65. Weis N, Stolz P, Krooss J, Meierhenrich U (1998) Gesundheitswesen 60:445
66. Thomas TC, Nishioka YA (1985) Bull Environ Contam Toxicol 35:460
67. Gebefügi I, Parlar H, Korte F (1979) Ecotoxicol Environ Saf 3:269
68. Yeboah PO, Kilgore WW (1984) Bull Environ Contam Toxicol 33:13
69. Warren JS, Lamparski LL, Johnson RL (1982) Bull Environ Contam Toxicol 29:719
70. Claas TJ, Kintrup J (1991) Fresenius J Anal Chem 340:446
71. Blessing R, Derra R (1992) Staub Reinh Luft 52:265
72. Lewis RG, MacLeod KE (1982) Anal Chem 54:310
73. Lewis RG, Jackson MD (1982) Anal Chem 54:592
74. Ball M, Hermann T, Wildeboer B, Koss G, Sagunski H, Czaplenski U (1993) Proceedings of indoor air '93, vol 2, p 201
75. Verein Deutscher Ingenieure (2000) Indoor air pollution measurement. Measurement of oentachlorophenol (PCP) and γ-hexachlorocyclohexane (lindane).GC/MS method. V4301, part 2
76. Verein Deutscher Ingenieure (2000) Indoor air pollution measurement. Measurement of pentachlorophenol (PCP) and γ-hexachlorocyclohexane (lindane).GC/ECD method. V4301, part 4 (draft)
77. Roper EM, Wright CG (1984) Bull Environ Contam Toxicol 33:476
78. Gebefügi I, Parlar H, Korte F (1976) Chemosphere 4:227
79. Dahms A, Metzner W (1979) Holz Roh-Werkst 37:341
80. Woiwode W, Wodarz R, Drysch K, Weichardt H (1980) Int Arch Occup Environ Health 45:153
81. Schnelle-Kreis J, Scherb H, Gebefügi I, Kettrup A, Weigelt E (2000) Sci Total Environ 256:125
82. Lane DA, Johnson ND, Hanley M-JJ, Schroeder WH, Ord DT (1992) Environ Sci Technol 26:126
83. US Environmental Protection Agency (1997) Exposure factors handbook. National Center for Environmental Assessment, Washington, DC
84. Verein Deutscher Ingenieure (2001) Measurement of indoor air pollution. Sampling of house dust. VDI 4300, part 8
85. Butte W, Heinzow B (2002) Rev Environ Contam Toxicol 175:1
86. Millette JR (2001) Microscope 49:201
87. Fergusson JE, Kim ND (1991) Sci Total Environ 100:125
88. Mølhave L, Schneider T, Kjaergaard SK, Larsen L, Norn S, Jorgensen O (2000) Atmos Environ 34:4767
89. Paustenbach DJ, Finley BL, Long TF (1997) Int J Toxicol 16:339
90. American Society for Testing and Materials (1994) Standard practice for collection of dust from carpeted floors for chemical analysis. ASTM method D 5439-94
91. American Society for Testing and Materials (1998) Standard practice for collection of dislodgeable residues from floors. ASTM method D 633-98
92. Macher JM (2001) Indoor Air 11:99
93. Colt JS, Zahm SH, Camann DE, Hartge P (1998) Environ Health Perspect 106:721

94. Rehwagen M, Rolle-Kampczyk U, Herbarth O, Schlink U, Krumbiegel P (1999) Gefahrst Reinh Luft 59:43
95. Rehwagen M, Rolle-Kampczyk U, Diez U, Borte M, Herbarth O (2000) Gefahrst Reinh Luft 60:305
96. Butte W, Hoffmann W, Hostrup O, Schmidt A, Walker G (2001) Gefahrst Reinh Luft 61:19
97. Salthammer T (2001) Umweltmed Forsch Prax 6:79
98. Santillo D, Johnston P, Bridgen K:(2001) The presence of brominated flame retardants and organotin compounds in dusts collected from parliament buildings from eight countries. Technical note 03/2001. Greenpeace Research Laboratories, Exeter, UK
99. Hong S, Kim J, Lemley AT, Obendorf SK, Hedge A (2001) J Chromatogr Sci 39:101
100. Reighard TS, Olesik SV (1997) Anal Chem 69:566
101. Reighard TS, Olesik SV (1996) Anal Chem 68:3612
102. Chuang JC, Callahan PJ, Menton RG, Gordon SM, Lewis RG, Wilson NK (1995) Environ Sci Technol 29:494
103. Moate TF, Furia M, Curl C, Muniz JF, Yu J, Fenske RA (2002) J Assoc Off Anal Chem Int 85:36
104. Chuang JC, Chou Y-L, Nishioka M, Andrews K, Pollard M, Menton R (1997) Field evaluation of screening techniques for polycyclic aromatic hydrocarbons, 2,4-diphenoxyacetic acid, and pentachlorophenol in air, house dust, soil and total diet. EPA/600/SR-97/109. Research Triangle Park
105. Chuang JC, Pollard MA, Chou YL, Menton RG, Wilson NK (1998) Sci Total Environ 224:189
106. Sagner G, Schöndube M (1982) In: Aurand K, Seifert B, Wegner J (eds) Luftqualität in Innenräumen. Fischer, Stuttgart, p 359
107. Petrowitz H-J (1986) Holz Roh-Werkst 44:341
108. Pluschke P (1996) Luftschadstoffe in Innenräumen. Springer, Berlin Heidelberg New York
109. Lemus R, Abdelghani A (2000) Rev Environ Health 15:421
110. Stolz P, Meierhenrich U, Krooss J, Weis N (1996) VDI-Berichte 1257:789
111. Butte W (1999) In: Salthammer T (ed) Organic indoor air pollutants. Wiley-VCH, Weinheim, p 171
112. Camann DE, Colt JS, Zuniga MM (2002) Proceeding of indoor air 2002, vol 4, p 860
113. Cizdziel JV, Hodge VF (2000) Microchem J 64:85
114. Schulz C, Becker K, Friedrich C, Krause C, Seifert B (1999) Proceedings of indoor air '99, vol 2, p 788
115. Bradman MA, Harnly ME, Draper W, Seidel S, Teran S, Wakeham D, Neutra R (1997) J Exposure Anal Environ Epidemiol 7:217
116. Gordon SM, Callahan PJ, Nishioka MG, Brinkman MC, O'Rourke MK, Lebowitz MD, Moschandreas MJ (1999) J Exposure Anal Environ Epidemiol 9:456
117. Lewis RG, Fortune CR, Willis RD, Camann DE, Antley JT (1999) Environ Health Perspect 107:721
118. Schomberg K, Winkens A (2000) Umweltmed Forsch Prax 5:331
119. Liebl B, Mayer R, Kaschube M, Wächter H (1996) Gesundheitswesen 58:332
120. Schmidt A, Hoffmann W, Hostrup O, Walker G, Butte W (2002) Gefahrst Reinh Luft 62:95
121. Krause C (1982) In: Aurand K, Seifert B, Wegner J (eds) Luftqualität in Innenräumen. Fischer, Stuttgart, p 309
122. Ott WR, Roberts JW (1998) Sci Am 278:72
123. Davis DL, Ahmed AK (1998) Environ Health Perspect 106:299
124. O'Rourke MK, Lizardi PS, Rogan SP, Freeman NC, Aguirree A, Saint CG (2000) J Exposure Anal Environ Epidemiol 10:672

125. Daniels JL, Olshan AF, Savitz DA (1997) Environ Health Perspect 105:1068
126. Infante-Rivard C, Labuda D, Krajinovic M, Sinnett D (1999) Epidemiol 10:481
127. Leiss JK, Savitz DA (1995) Am J Public Health 85:249
128. Meinert R, Kaatsch P, Kaletsch U, Krummenauer F, Miesner A, Michaelis J (1996) Eur J Cancer 32A:1943
129. Meinert R, Schuz J, Kaletsch U, Kaatsch P, Michaelis J (2000) Am J Epidemiol 151:639
130. Zahm SH, Ward MH (1998) Environ Health Perspect 106(Suppl 3):893
131. Guilette LJ, Meza MM, Aquilar MG, Soto AD, Garcia IE (1998) Environ Health Perspect 106:347
132. Daniels JL, Olshan AF, Teschke K, Hertz-Picciotto I, Savitz DA, Blatt J, Bondy ML, Neglia JP, Pollock BH, Cohn SL, Look AT, Seeger RC, Castleberry RP (2001) Epidemiol 12:20
133. Olshan AF, Daniels JL (2000) Am J Epidemiol 151:647
134. Gurunathan S, Robson M, Freeman N, Buckley B, Roy A, Meyer R, Bukowski J, Lioy PJ, (1998) Environ Health Perspect 106:9
135. Buckley TJ, Liddle J, Ashley DL, Paschal DC, Burse VW, Needham LL, Akland G (1997) Environ Int 23:705
136. Krause C, Englert N (1980) Holz Roh-Werkst 38:429
137. Meissner T, Schweinsberg F (1996) Toxicol Lett 88:237
138. Butte W, Walker G, Heinzow B (1998) Umweltmed Forsch Prax 3:21
139. Moschandreas DJ, Karuchit S, Kim Y, Ari H, Lebowitz MD, O'Rourke MK, Gordon S, Robertson G (2001) J Exposure Anal Environ Epidemiol 11:56
140. Lioy PJ, Edwards RD, Freeman N, Gurunathan S, Pellizzari E, Adgate JL, Quackenboss J, Sexton K (2000) J Exposure Anal Environ Epidemiol 10:327
141. Tulve,NS, Suggs JC, McCurdy T, Cohen Hubal EA, Moya J (2002) J Exposure Anal Environ Epidemiol 12:259
142. Calabrese EJ Stanek E (1991) Regul Toxicol Pharmacol 13:278
143. Arbeitsgemeinschaft der leitenden Medizinalbeamtinnen und -beamten der Länder (1985) Standards zur Expositionsabschätzung. Behörde für Arbeit, Gesundheit und Soziales, Hamburg
144. Lewis RG, Fortune CR, Blanchard FT, Camann DE (2001) J Air Waste Mangem Assoc 51:339
145. Ewers U, Krause C, Schulz C, Wilhelm M (1999) Int Arch Occup Environ Health 72:255
146. Friedrich C, Helm D, Becker K, Hoffmann K, Krause C, Nöllke P, Schulz C, Seiwert M, Seifert B (2001) Umwelt-Survey 1990/92, Band VI, Hausstaub. Deskription der Spuren-element- und Biozidgehalte im Hausstaub der Bundesrepublik Deutschland. Umwelt-bundesamt, Berlin
147. Lebowitz MD, O'Rourke MK, Gordon S, Moschandreas DJ, Buckley T, Nishioka M (1995) J Exposure Anal Environ Epidemiol 5:297
148. O'Rourke MK, Van de Water PK, Jin S, Rogan SP, Weiss AD, Gordon S (1999) J Exposure Anal Envrion Epidemiol 9:435
149. Robertson GL, Lebowitz MD, O'Rourke MK, Gordon S, Moschandreas D (1999) J Exposure Anal Environ Epidemiol 9:427
150. Adgate JL, Clayton CA. Quackenboss JJ, Thomas KW, Whitmore RW, Pellizzari ED, Lioy PJ, Shubat P, Stroebel C, Freeman NC, Sexton K (2000) J Exposure Anal Environ Epidemiol 10:650
151. Quackenboss JJ, Pellizzari ED, Shubat P, Whitmore RW, Adgate JL, Thomas KW, Freeman NCG, Stroebel C, Lioy PJ, Clayton AC, Sexton K (2000) J Exposure Anal Environ Epidemiol 10:145
152. Ad-hoc-Arbeitsgruppe aus Mitgliedern der Innenraumlufthygiene-Kommission (IRK) des Umweltbundesamtes und des Ausschusses für Umwelthygiene der AGLMB (1997) Bundesgesundheitsblatt 40:234

153. Ad-hoc-Arbeitsgruppe aus Mitgliedern der Innenraumlufthygiene-Kommission (IRK) des Umweltbundesamtes und des Ausschusses für Umwelthygiene der AGLMB (1996) Bundesgesundheitsblatt 39:422

154. Seifert B, Englert N, Sagunski H, Witten J(1999) Proceedings of indoor air '99, vol 1, p 499

155. Pluschke P (1999) In: Salthammer T (ed) Organic indoor air pollutants. Wiley-VCH, Weinheim, p 292

156. US Environmental Protection Agency (2001) Integrated risk information system (IRIS on line). Office of Health and Environmental Assessment, Cincinnati, OH

157. Louis JB, Kisselbach KC Jr (1987) Bull Environ Contam Toxicol 39:911

158. Lederer P, Angerer J (1997) Umweltmed Forsch Prax 2:32

159. Fromme H (1991) Öff Gesundh-Wes 53:661

160. Stolz P, Meierhenrichs U, Krooss J (1994) Staub Reinh Luft 54:379

161. Haumann T, Thumalla J (2002) Proceeding of indoor air 2002, vol 4, p 865

The Handbook of Environmental Chemistry Vol. 4, Part F (2004): 117–147
DOI 10.1007/b94833
© Springer-Verlag Berlin Heidelberg 2004

Indoor Particles, Combustion Products and Fibres

Lidia Morawska (✉)

International Laboratory for Air Quality and Health, Queensland University
of Technology, 2 George Street, Brisbane, QLD 4001, Australia
l.morawska@qut.edu.au

Abstract Pollutants in an indoor environment are a complex mixture of gases, vapours and particles in either the liquid or the solid phase, suspended in the air, settled or adsorbed on or attached to indoor surfaces. The pollutants originate from a multiplicity of indoor and outdoor sources. The pollutant mixture is dynamic, involved in numerous physical and chemical processes and changes its characteristics with time. Its composition and concentration depend on the strengths of indoor sources, the concentration of pollutants outside and the properties of heating-ventilation and air-conditioning systems. The spatial distribution of pollutant concentration within an indoor environment is often inhomogeneous. Particulate matter in an indoor environment includes particles which are airborne as well as those which are settled on indoor surfaces: dust. The particles vary in chemical properties, which depend on the origin of the particles and differ for particles in different size ranges. The particles can, for example, be combustion products, dust or bioaerosols, and can act as carriers of adsorbed chemicals, biocontaminants or condensed gases. Particles are a key component of emissions from all the combustion sources. In particular, a significant indoor combustion product, environmental tobacco smoke is a mixture of particle and gaseous products of smoke exhaled into the air by smokers and is mixed with the smoke resulting from smouldering of a cigarette between puffs. This chapter is focused on particulate matter, its origin, characteristics and behaviour in an indoor environment. In addition, several important classes of indoor pollutants are discussed: those which are entirely or partially composed of particulate matter. These include environmental tobacco smoke and combustion products from other sources, such as wood smoke or vehicle emissions, and also fibres, in particular, asbestos.

Keywords Indoor air pollution · Indoor particles · Environmental tobacco smoke · Fibres · Dust

1
Origins of Indoor Airborne Particles

A large number of sources contribute to ambient airborne particulate matter and include motor vehicles, power plants, wind-blown dust, photochemical processes, cigarette smoking and nearby quarry operation. Particles encountered in indoor air can be generated from either indoor or outdoor sources. The indoor sources of particles include

- Occupants (which are humans and pets contributing to air pollution through functions natural to life processes or through activities conducted).
- Soil, water.
- Cooking.
- Tobacco combustion.
- Combustion appliances.
- Building materials.
- Furnishings.
- Consumer products.
- Maintenance products.

There is a significant variation between particles generated not only by different sources, but even by the same type of source. The most significant indoor sources are smoking, cooking and occupant movement.

Some of the particles present in the air are primary and some are secondary in nature. A primary particle is a particle present in the air in the form in which it was generated by a source, while a secondary particle is formed in the air by gas-to-particle conversion. An important characteristic of airborne particles is their size distribution. The size of the particles strongly affects particle behaviour and fate in atmospheric systems as well as deposition in the human respiratory tract. The size distribution is also the main factor in choosing the instrumentation to be used for particle detection. Particles in supermicrometre-size ranges (larger than 1 μm) are usually primary in nature and result mainly from mechanical processes such as cleaning and physical activity of the occupants indoors. Particles in submicrometre ranges are generated mainly from combustion processes as well as from secondary processes such as gas-to-particle conversion and nucleation or photochemical processes. In an indoor environment the main sources of submicrometre particles include smoking, cooking (particularly frying and broiling), operation of gas burners, gas ovens and electric toasters [1].

A source signature or fingerprint of a source consists of the physical and/or chemical characteristics of the emissions, which are specific and at best unique for that source. The availability of such signatures is very important for the determination of absolute and fractional contributions from specific sources (source apportionment), and in turn for developing emission inventories. Quantitative source apportionment is a very complex undertaking as, on one hand, ambient air contains a dynamic mixture of pollutants emitted from various sources. This mixture undergoes continuous change with time as the interactions between pollutants take place and as the components of the mixture are removed from the air owing to the presence of various sinks. On the other hand, it is only rarely that specific emission characteristics are unique to a particular source. More often emissions from other sources display some of these characteristics as well. The following are used as source signatures:

- Physical aspects of particles (number or mass size distribution, density and shape).
- Chemical aspects of signatures: elemental composition, elemental ratios, characterisation of chemical form, isotope ratios, organic compounds.

In addition to the source signatures, certain elements or compounds have been used as markers of emissions from specific sources [2]. A suitable marker should be (1) unique or nearly unique to the emissions from the source under consideration, (2) similar in emission rates for a variety of the same type of fuels, (3) easily detected in air at low concentrations and (4) present in consistent proportion to compounds that have effects on human health. Additionally, an ideal marker should be easily (in real-time), accurately and cost-effectively measurable.

Quantitative measures of emissions are emissions factors and emission rates. A source emission factor is typically defined as the amount of a chemical species, mass, particle number, etc. emitted per unit mass of fuel burned or per defined task performed [3]. The former is often referred to as a mass-based emission factor and has a unit such as grams per kilogram. The latter can be called a task-based emission factor. The unit of the task-based emission factor depends on the definition of the tasks. For example, a task can be the number of cigarettes smoked or a certain distance driven by a motor vehicle and thus the units may be grams per cigarette or grams per kilometre, respectively. The emission rate, on the other hand, is the amount of a chemical species, mass, particle number, etc. emitted by the source per unit time. For example, emissions from stoves are usually characterised in terms of emission factors. Similarly, re-entrainment of settled dust to the air is represented by resuspension rates. Emission factors and emission rates vary significantly not only between different types of sources but also between sources of the same type.

Certain sources, for example, cooking or smoking, or activities such as walking are always associated with relatively high emission factors or rates, and thus always contribute in a measurable way to indoor concentration levels of particles [4]. For other sources or activities, however, for example, cleaning or vacuuming, conflicting results have been reported in the literature. While a number of studies, for example, found cleaning to contribute to increase of particle concentration, the PTEAM study [5] determined the contribution from cleaning to be not statistically significant. The reasons for the conflicting results relate, on one hand, to varying emission rates from these sources, which are sometimes very low, and also to the degree of dilution of particles introduced from these sources to indoor environments, which depend on air exchange rates. The combination of low emission factors or rates and high exchange rates can result in insignificant contribution from certain sources in certain environments. It is important at keep in mind that the choice of measuring method is also a factor of significance: if an inappropriate method were chosen for detection of the particles emitted, the conclusions about emissions from the sources would be erroneous.

The concentration of particles as well as of other pollutants in an indoor environment depends on a number of factors:

- The type, nature and number of sources.
- Source use characteristics.
- Building characteristics.
- Outdoor concentration of pollutants.
- Infiltration or ventilation rates.
- Air mixing.
- Removal rates by surfaces, chemical transformation or radioactive decay.
- Existence and effectiveness of air-contaminant removal systems.
- Meteorological conditions.

The role of the outdoor air on the indoor particle characteristics cannot be overestimated. In the absence of active indoor sources particles generated by outdoor sources which penetrated indoors are the main constituent of indoor particles. In a typical outdoor urban environment, motor vehicle emissions constitute the most important source of all pollutants including particles. The emissions from motor vehicles penetrate indoors and their concentration in indoor air is often comparable to the concentration outdoors.

The relative importance of indoor and outdoor sources depends on the environment and lifestyle of the occupants. For example, there will be little contribution from indoor combustion sources in an indoor environment which does not require heating, where cooking is conducted using electric stoves, and where there are no smokers. On the other hand, environments with operating open fires or where cigarettes are smoked inside could have concentrations which are orders of magnitude higher than in an outdoor environment.

Open-fire burning presents a particularly severe problem in developing countries, where in many places it is the most affordable or the only available way of cooking or heating.

2
Physical, Chemical and Biological Properties of Particles

Airborne particles can be classified and characterised in a number of ways and, for example according to their physical, chemical or biological properties. Also, many different terms are used in relation to airborne particles. Some of them identify particles by their sizes, others by the processes which led to their generation and some by the ability of the particles to enter the human respiratory tract [6, 7]. In particular, an aerosol is an assembly of liquid or solid particles suspended in a gaseous medium long enough to enable observation or measurement, and a particle or particulate is a small, discrete object.

The most important physical properties of aerosol particles include the number and number size distribution, the mass and mass size distribution, the surface area, the shape and the electrical charge. To a larger extent these are the physical properties of particles which underlie particle behaviour in the air and ultimately removal from atmospheric systems. Of particular importance is the size of the particles. The efficiency of various forces acting on particles and the processes to which they are subjected in the air depends strongly on particle physical properties, of which size is one of the most important. Health and environmental effects of particles are strongly linked to particle size, as it is the size which is a predictor of the region in the lung where the particles would deposit or the outdoor and indoor locations to which the particles can penetrate or be transported. Also sampling of particles and choice of appropriate instrumentation and methodology is primarily based on particle physical properties.

2.1
Particle Size and Its Relation to Formation Mechanisms

Various classifications and terminologies have been used to define particle size ranges. The division most commonly used is between fine and coarse particles, with the boundary between these two fractions widely accepted as 2.5 μm. However, this division has been defined differently by different authors, in relation to different aerosols and for different applications, and ranges from 1 to 2.5 μm. The division line often used in aerosol science and technology is somewhere between 1 and 2 μm [6–8]. The rationale behind this is that this is the range of a natural division between smaller particles, which are generated mainly from combustion and other process leading to gas-to-particle conversion, and larger particles, which are generated from mechanical processes.

Ultrafine particles have been defined as those, which are smaller than 0.1 μm. Another classification is into submicrometre particles, which are smaller than 1 μm, and supermicrometre particles, which are larger than 1 μm. The terminology that has been used in the wording of the ambient air quality standards, and also for characterisation of indoor and outdoor particle mass concentrations, includes $PM_{2.5}$ and PM_{10} fractions and the total suspended particulate (TSP). $PM_{2.5}$ (fine particles) and PM_{10} are the mass concentrations of particles with aerodynamic diameters smaller than 2.5 and 10 μm, respectively (more precisely the definitions specify the inlet cutoffs for which 50% efficiency is obtained for these sizes). TSP is the mass concentration of all particles suspended in the air. There have been references made in the literature to PM_1 or $PM_{0.1}$ fractions, which imply mass concentrations of particles smaller than 1 and 0.1 μm, respectively. These terms should be used with caution, as particles below 1 μm, and even more those below 0.1 μm, are more commonly measured in terms of their number rather than their mass concentrations, and therefore these terms could be misleading.

It should be kept in mind that the divisions between the different particle size classes are somewhat arbitrary. On the one hand, there are no natural boundaries between these size classes as nature itself does not provide a perfect division. On the other hand, all natural sources (versus laboratory sources) produce particles within a certain range of diameters (polydisperse particles). Therefore, there is no sharp boundary delineating the contribution of particles from a given particle source. This argument also extends to effects produced by the particles. For example, it cannot be expected that there will be much difference between 0.09 and 0.11-μm particles in terms of their composition and behaviour in atmospheric systems, nor in their penetration into the lung or the health effects they cause, despite the second particle being outside the defined ultrafine range.

While the previous classification of particles considers only their sizes, as already discussed, the particle size is a consequence of the process which led to the generation of the particle, and thus is also dependent on the source. Submicrometre particles are generated mainly from combustion, gas-to-particle conversion, nucleation processes or photochemical processes, while larger particles result mainly from mechanical processes, for example, cutting, grinding, breaking and wear of material and dust resuspension. Particles in the submicrometre range typically contain a mixture of components including soot, acid condensates, sulfates and nitrates, as well as trace metals and other toxins. Coarse particles contain largely earth crustal elements and compounds.

Particle size distribution can also be described in terms of modes which correspond to peaks within the distribution. This classification relates to particle formation mechanisms; however, it also implies particle size ranges. The location of the modes is variable, depending on the specific sources and other local atmospheric conditions. Particles can be classified into the following modes:

- Nuclei mode: particles in this mode are formed by nucleation of atmospheric gases in a supersaturated atmosphere and their size is of the order of nanometres.
- Accumulation mode: particles in this mode originate from primary emissions as well as through gas-to-particle conversion, chemical reactions, condensation and coagulation.
- Coarse mode: particles generated by mechanical processes.

2.2
Particle Shape and Equivalent Diameter

Particles vary significantly in shape, which in general relates to the particle formation process or their origin. Some particle shapes are fairly simple and regular; however, the majority are of a varying degree of irregularity or complexity. Particles resulting from coagulation and agglomeration of smaller solid particles are usually highly irregular and display fractal properties. Examples of these are particles resulting from combustion process such as vehicle emissions or tobacco smoking, and they are agglomerates of carbonaceous particles. Similarly dust particles and particles resulting from mechanical breaking, grinding, etc. have generally irregular shapes. By contrast, liquid aerosol particles are usually spherical, while simple fibres are rod-shaped. Biological particles are of complex shapes and differ significantly between each other. Microscopic images of particles of various shapes collected inside and in the vicinity of residential houses in Brisbane, Australia, are presented in Fig. 1.

For practical applications particles of complex or irregular shapes are usually characterised by only one or two parameters: those which can be measured. These are most commonly particle diameter and for fibres, their lengths and width. The diameter is a characteristic of spherical objects; however, as already explained only a small fraction of airborne particles are spherical. A means of representation of particle irregular shapes is by particle equivalent diameter, which is the diameter of a sphere having the same value of a physical property as the irregularly or complex shaped particle being measured. The equivalent diameter relates to particle behaviour (such as inertia, electrical or magnetic mobility, light scattering, radioactivity or Brownian motion) or to particle properties (such as chemical or elemental concentration, cross-sectional area, volume-to-surface ratio). Therefore the particle diameter determined experimentally depends on the choice of the particle properties or the behaviour measured and thus application of different methods for measurements of the particle diameter usually result in somewhat different values of the diameter.

The most commonly used equivalent diameters are aerodynamic (mainly for particles larger than 0.5 μm), diffusion (for particles smaller than 0.5 μm), light scattering (for various ranges from about 0.1 μm and larger), and mass equivalent (mainly for larger particles). The aerodynamic (equivalent) diam-

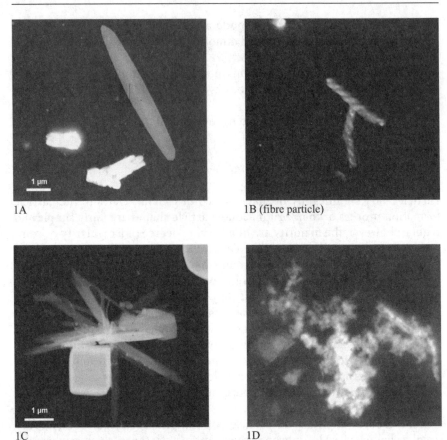

1A

1B (fibre particle)

1C

1D

Fig. 1a–d Particles collected in and outside residential houses in Brisbane and examined with an energy-dispersive X-ray analyser attached to a transmission electron microscope. **a** There are three big particles: two *lighter particles*, with dominant elements S, Ca, O and Mg, and one *darker particle*, with dominant elements Ca, S, Na, O and Mg. **b** There are two big particles, with dominant elements C, Cl, Na, Mg, K and O. The particles are probably fine pieces of insect body or plant material. **c** There are two types of particles, a *square particle* (NaCl crystal) and many *big fibrous particles*, with dominant elements Ca, S, Na, O and Mg. **d** There are many particles jointed together in this picture, with dominant elements Fe, Pb and Si. The particles are probably from vehicle emissions

eter, D_a, is the diameter of a unit-density sphere having the same gravitational settling velocity as the particle being measured. D_a is given by Eq. (1):

$$D_a = D_p k \sqrt{\frac{\rho_P}{\rho_0}}, \tag{1}$$

where D_p is the physical particle diameter, ρ_P is the density of the particle, ρ_0 is the reference density (1 g cm^{-3}) and k is a correction factor, which can be approximated by $k \approx 1$ for many applications.

An important feature of many types of particles, particularly those of complex shapes resulting from combustion processes, is their fractal structure. For example Schmidt-Ott [9] measured fractal dimensions of ultrafine particles resulting from gas-to-particle conversion and showed that such particles have very large surface areas. Thus, large quantities of chemicals can be adsorbed by such particles and in such cases knowledge of the aerodynamic diameter alone provides no information on the possible amount of adsorbed compounds. Understanding of fractal objects and their links and relationship to various processes in nature can help in understanding the structure and growth of particles [10, 11].

2.3
Particle Sizes and Size Distribution

The smallest and the largest airborne particles can differ in size by up to 5 orders of magnitude and range from about 1 nm to about 100 µm. The former is molecular size and the later is the size above which particles sediment rapidly owing to gravitational force. Almost all of the sources generate particles with some distribution of sizes, a so-called polydisperse aerosol, rather than particles of a single size, a monodisperse aerosol. The spread of the particle size distribution is characterised by an arithmetic or a geometric (logarithmic) standard deviation. The most common ways of characterisation of a particle distribution are in terms of its mean size, which is the average of all sizes, the median size, which means an equal number of particles above and below this size, or the mode size, which is the size with the maximum number of particles. The terms used include count, number or mass median diameter, which are abbreviated as CMD, NMD or MMD, respectively. MMAD is mass median aerodynamic diameter.

The size distribution of particles generated by most sources is lognormal, which means that the particle concentration versus particle size curve is "normal" (bell-shaped) when the particles are plotted on a logarithmic scale. The width of the peak in the distribution is characterised by the geometric standard deviation. Different emission sources are characterised by different size distributions. These distributions are not unique to these particle sources alone; however, the information on the size distribution can help to identify the source contribution to particle concentrations in ambient air, and also serve as a source signature. In general, submicrometre and supermicrometre particles result from different generation processes and only occasionally the same source generates particles with a broad size distribution, covering both fine and coarse ranges.

Numerous researchers have presented information about size distributions of particles originating from various sources. Wallace [12] reported that at the peak of the number concentrations the diameters of the particles produced by several sources are 0.01–0.02 µm for gas burners, a gas oven and a toaster, about 0.1 µm for incense, 0.05–0.1 µm for frying and broiling and 5–10 µm for walking, moving one's arms and even sitting in front of a computer. Abt et al. [13] showed that cooking (including broiling and baking, toasting and barbecuing) produced particles with a volume diameter in a size range between 0.13 and 0.25 µm, cleaning and smoking particles smaller than 1 µm, while moving of people and sautéing resulted in particles in the range between 3 and 4.3 µm. Frying was associated with both fine and coarse particles. Measurements conducted in an apartment in Taipei by Li et al. [14, 15] showed that the NMDs of particles originating from the background, smouldering cigarettes, burning mosquito coils and joss sticks were 0.07, 0.085, 0.08 and 0.07 µm, respectively. In a study conducted by Kleeman et al. [16] natural gas, propane and candle flames generated particles between 0.01 and 0.1 µm and meat charbroiling showed a major peak in the particle mass distribution at 0.1–0.2 µm with some material present at larger particle sizes but not above 1.0 µm. Biological particles are another class which include many different species varying not only in their biological composition but also in their size. In general viruses range from 0.02 to 0.3 µm, bacteria from 0.5 to 10 µm, fungi from 0.5 to 30 µm, pollen 10 to 100 µm and house dust mites are about 10 µm.

Particles in ambient air constitute a mixture originating from different sources, and thus the individual components of the mixture are characterised by different size distributions. The measured distribution of this mixture may or may not display individual peaks from the contributing sources, and thus may or may not be used for source identification. In many cases, however, the characteristics of the size distribution can be a useful tool in source characterisation.

Concentration of particles in the air as well as particle size distributions can be considered either in terms of particle number or mass. In terms of number, the vast majority of airborne particles are in the ultrafine range. For example, in urban outdoor air where motor vehicle emissions are a dominant pollution source, over 80% of particulate matter in terms of number is in the ultrafine range [17]. Since outdoor particles contribute significantly to indoor particle concentrations, also in indoor air particle number concentration is usually dominated by the smallest particles. However, most of the mass of airborne particles is associated with large particles since the mass of ultrafine particles is often very small in comparison with the mass of larger particles. The particle surface area in turn is largest for particles somewhat above the ultrafine size range.

The relationship between particle number, the surface area and the volume is presented in Fig. 2, using as an example a typical urban air particle size distribution measured in Brisbane [18]. This relationship was derived using the measured particle number size distribution and calculating the particle surface and volume distribution (assuming their sphericity) and by plotting $dN/dlogD_P$, $dA/dlogD_P$ and $dV/dlogD_P$, which represent particle number, surface area and volume, respectively, per logarithmic size interval. The particle mass can be calculated from the volume when the particle density is known or can be assumed. It can be seen from Fig. 2 that the peak in the number

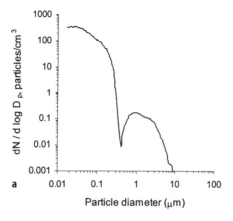

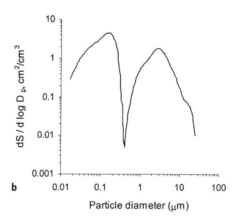

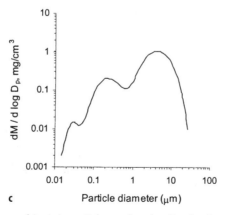

Fig. 2a–c Typical urban ambient air particle number size distribution measured in Brisbane [8], and calculated from **a** the number distribution, **b** the surface area distribution and **c** the volume size distribution, respectively

distribution spectrum appears in the area where there is almost no volume in the volume distribution spectrum and vice versa, the peak in the volume distribution spectrum is where the particle number is very low.

Since different sources contribute to the generation of particles in the sub-micrometre range, which is predominant in the particle number, and different sources to larger particles, which predominate in the mass, it is only occasio-nally that there is a correlation between the particle number and the mass. In general, however, only limited information, or no information at all, can be obtained about the particle number from the measurements of the particle mass, and vice versa. The degree of correlation depends on specific local con-ditions, and in particular better correlations are achieved for conditions when the majority of particles in the fine and coarse size ranges are related to the same generation process.

Comparison between different particle mass fractions, for example, $PM_{2.5}$ and PM_{10}, reveals that there is often a relatively high degree of correlation be-tween the fractions. One of the main reasons for this is that $PM_{2.5}$ is a fraction of PM_{10}, often quite significant, and for sources operating in a stable manner, the relation between $PM_{2.5}$ and PM_{10} emissions remains more or less constant. Outdoor ratios of particle mass fractions could be indicators of these ratios in an indoor environment where natural ventilation is used and in the absence of indoor sources. In the presence of operating indoor sources, however, the indoor and outdoor ratios could differ substantially. For example, Chao et al. [19] reported that the outdoor PM_{10} in Hong Kong was around 50–70% of the TSP level; however, indoor ratios of PM_{10} to TSP measured in eight residential premises varied from 81.9% to 97.7%.

2.4
Chemical Composition of Airborne Particles

The chemical composition of indoor particles is very complex and depends on the particle origin as well as on any postformation processes in which the particle is involved. Some types of particles, like asbestos and glass fibres, consist of inorganic materials, while other types, like cellulose fibres, are purely organic. The most important chemical properties of particles include

- Elemental composition.
- Inorganic ions.
- Carbonaceous compounds (organic and elemental carbon).

Each of these consists of about one third of the total aerosol mass and has an impact on particle proprieties and behaviour, such as adsorption/desorption, water solubility or extractability.

Interest in different aspects of particle chemical compositions is driven either by the risk associated with certain components or with the potential for application of the knowledge about the components towards some practical purposes. For example, interest in elemental composition derives from the potential health effects of heavy elements like lead, arsenic, mercury and cadmium, and the possibility of using the elements as source tracers [4]. Water-soluble ions such as potassium, sodium, calcium, phosphates, sulfates, ammonium and nitrate associate themselves with water in indoor environments and can also be used for source apportionment. Carbonaceous compounds are composed of organic and elemental carbon. The former can contain a wide range of compounds such as polycyclic aromatic hydrocarbons, pesticides, phthalates, flame retardants and carboxylic acids, some of which are tracers for certain sources, while the latter is sometimes termed "soot", "black carbon" and "graphitic carbon".

Owing to their irregular shapes the majority of the particles present in an indoor environment have large surface areas, which provide an opportunity for particles to serve as sinks for a variety of organic species. Semivolatile substances are found both in the particulate and in the vapour phases, volatile compounds occur mostly in the gas phase and substances with very low vapour pressure are adsorbed almost exclusively. In addition to the surface area of the particles, the vapour pressure of organic compounds also plays an important role in determining whether they will be found in the gas phase or adsorbed on particle matter. More information on particle chemistry is provided in Sect. 4.3.

2.5
Bioaerosols

A certain fraction of particles in indoor and outdoor air is of biological origin. In addition to particles, some volatile organic compounds are also of microbial origin (MVOC). According to the definition formulated at the IGAP workshop in Geneva in June 1993, "Biological Aerosol Particles (BAP) describe airborne solid particles (dead or alive) that are or were derived from living organism, including micro-organisms and fragments of varieties of living things" [20]. The sources of biological particles can be classified as (1) animal or human sources, (2) terrestrial including rural (plants) and urban sources, (3) aquatic sources and (4) atmospheric sources. Biological particles in indoor environments include viruses; bacteria, which at cell destruction release endotoxins; animal dander (cats, dogs, rabbits, rodents, birds), which contains allergens (e.g. cats Fel d1); cockroaches and other insects, which contain allergens; mites, which contain allergens (e.g. Der p1) and release faeces; moulds (filamentous fungi), which release mould spores (which in turn when damaged break into spore fragments which include allergens and contain β-glucans) and release primary metabolites (MVOC) and secondary metabolites (mycotoxins) and pollens [4].

Various units are used in relation to the concentration of biological particles in the air and in particular:

1. For viable microorganisms: colony forming units (per cubic metre) for bacteria and fungi and plaque forming units (per cubic metre) for viruses.

2. For nonviable microorganisms: the number of individual microorganisms per cubic metre, the number of microorganism-containing particles per cubic metre, or micrograms per cubic metre.

The presence and distribution of biological particles varies significantly between different indoor environments. The particles can be suspended in the air, attached to indoor surfaces, to the dust accumulated in the building and also present in any internal parts of the building structure or its operating systems (inside walls, air-conditioning units, ducts, etc.). The concentration levels of the particles can vary by orders of magnitude and depend on a range of local conditions and factors affecting their growth, survival, transport and removal from the air. Most commonly the concentration of biological particles is significantly lower than the concentration of nonbiological particles, with concentrations of the former being in the range 10^1–10^4 m^{-3} and the latter in the range 10^9–10^{11} and 10^6–10^7 m^{-3} for particles in submicrometre and supermicrometre ranges, respectively [4].

3
Transport and Behaviour

Airborne particles are subjected to a multiplicity of processes, interactions and reactions which change the characteristics the particles initially had when introduced into the air. The most important processes include sedimentation, deposition on surfaces, coagulation and changes by evaporation or condensation. Some of the emission products undergo rapid changes, for example, combustion-related, which are reactive mixtures of hot gases and particles, while others, like mechanical dust, are less so. Particles generated indoors and measured some time after emission often have different characteristics from those measured immediately after formation. The residence time of particles in the air depends on the nature of the processes they are involved in, and varies in indoor air from seconds to minutes or hours. Larger particles (of a micrometre size range and more in aerodynamic diameter) are removed from the air mainly through gravitational settling (with particles above 100 μm settling almost immediately after becoming airborne), while smaller particles are removed by precipitation or diffusional deposition.

The processes of the highest significance in affecting indoor particle concentration levels and other characteristics include

- Penetration of outdoor particles indoors through open doors, windows as well as through the building envelope. A measure of the ability of the particles to penetrate the building envelope is defined as the penetration factor [21] (sometimes also called the penetration coefficient). In cases when windows or doors of a building are open, they provide the main penetration route for the particles, and the relative importance of penetration through the building envelope becomes insignificant. Under such circumstances the penetration factor for particles in all size ranges is very close to unity, which means that particles enter buildings very easily with the air which carries them. Such situations are characterised by large air exchange rates. Recent studies conducted by Mosley et al. [22] and Long et al. [23] showed that the penetration factor is generally lower than unity for situations when the particles travel a torturous path through a building envelope, and also that it is strongly size dependent.

- Deposition of particles on indoor surfaces. Fundamentals of the theory of particle deposition are presented in Ref. [7] and describe the process of particle flux towards the surface due to the gradient in the particle concentration established in the region of the surface. At the surface the concentration is close to zero owing to the deposition and increases at a distance away from the surface (e.g. a wall) to achieve equilibrium usually a few centimetres from the surface. The deposition rate is defined as the number of particles depositing per unit surface area per unit time (per square metre per second). Particle deposition on indoor surfaces strongly depends on particle size and is governed by the processes of particle diffusion towards the surfaces, which is of particular significance for very small particles, and of gravitational sedimentation, which is significant for larger particles. In addition, airflows induced by convection currents or the action of fans, as well as air turbulence, can increase particle transport towards the surface and thus the deposition. Deposition is also dependent on the surface areas and on surface characteristics, with sticky surfaces resulting in more deposition and smoother ones in less deposition. The larger the surface area, the higher the probability of particle deposition, and therefore furnished rooms, with lots of surface area, will have a higher deposition rate than bare rooms. Additional factors affecting particle deposition are the presence of surface charge, which increases the deposition rate, temperature gradient, which results in convective currents and thermophoretic deposition, and room volume. This multiplicity of factors affecting particle deposition results in deposition being a highly variable process, site-specific and difficult to quantify either through experimental studies or modelling. Many studies have shown that there is a considerable difficulty associated with decoupling and quantifying separately particle deposition and other parameters describing particle dynamics, for example, the penetration factor [23].

- Resuspension of particles deposited on surfaces. Particles that have deposited on indoor surfaces, settled dust, may be resuspended from the surfaces and re-entrained into the air. To resuspend the particles, a certain force must be applied and energy used to detach the particles from the surface. According to the theory of aerosol interactions [6], while most adhesion forces are linearly dependent on the particle diameter, most detachment forces are proportional to the particle diameter to the second or third power. The differences between adhesion and detachment forces in their dependence on the particle diameter result in large particles being more readily detached than small ones. The theory was confirmed in a number of field studies and, for example, it was concluded in Ref. [24] that submicrometre particles are essentially non-resuspendable under circumstances encountered in residences. The resuspension rate was shown to increase as the particle size increases, and also similarly to several other studies showed that all the normal activities of the occupants, like walking (even walking in and out of the room), moving around, children playing, or cleaning, result in an increase in the number of supermicrometre particles.

– Removal of particles from an indoor environment by ventilation and filtration. As a result of the operation of natural or mechanical ventilation systems particles generated indoors are removed from indoor environments, but at the same time particles from outdoors penetrate indoors. Filers remove a certain fraction of particles from the air supplied indoors. Ventilation and air filtration are the main remedial actions available to reduce the concentration of airborne contaminants indoors. The design and operation of the filtration and ventilation systems and the type of filters used are of critical importance.

Generally, ventilation represents a dilution control of indoor pollution, which means that contaminated indoor air is diluted or displaced with "clean" outdoor air. The ventilation performance depends on room geometry, the ventilation method applied and on the operating conditions, as well as on the location and strength of the sources and the types of contaminants generated [25]. The ventilation rate or the air-exchange rate is defined as the ratio between the outdoor air flow rate and the effective volume of ventilated space. The air-exchange rate varies and depends on the climate, the type of building and its operation as well as the lifestyle of the occupants. For example, as summarised in Ref. [26] in residential houses in Australia, with a mild-to-warm climate, air exchange was reported to be relatively high, reaching on average 26.3 h^{-1}, while in Canada and Sweden, of much cooler climates, it was reported to be 4.4 and 3.7 h^{-1}, respectively The ventilation rate in office buildings is usually much lower, and in the USA was found to be 0.9 h^{-1} and in Brisbane about 0.8 h^{-1}.

Mechanical filtration systems are intended to limit the introduction of pollutants from outdoors to indoors. The efficiency of such systems generally depends on the filter properties and the aerodynamic properties of filtered particles [26]. The efficiency of filters varies from 5% to 40% for low-efficiency filters, such as dry media filters, panel and bag filters, from 60% to 90% for electrostatic precipitators to over 99% for high-efficiency particulate air filters. Not only the filters, but the whole heating, ventilation and air-conditioning system contributes to particle reduction, owing to particle losses on the cooling/heating coil and other parts of the system. The selection of a system depends on the type of indoor environment, outdoor and indoor sources, the demand on the level of reduction of pollutant concentrations and the cost associated with purchase, operation and maintenance of the system.

- Chemical reactions involving vapours and gases leading to particle generation. Examples of such processes are reactions between ozone and various terpenes in indoor environments, which have been shown to result in a significant increase in the number and mass concentrations of submicrometre particles [27]. For example, 20–40 ppb O_3 and several hundred parts per billion of terpenes can generate an additional 5–40 μg m^{-3} of fine particles [28]. Initially more than half of the mass increase is in the ultrafine fraction; however, the particle number and mass size distribution evolves over hours, shifting the peak of the distribution towards larger sizes. The products of ozone/terpene reactions include hydroxyl radicals that can, in turn, react with other indoor organic compounds such as toluene to produce low-volatility products, which further contribute to particle formation. The reaction of ozone with isoprene produces much less particulate matter than the reactions involving terpenes. This is related to the lower molecular weight of isoprene (half the weight of terpene) and also to the differences in volatility of the major products. In general, more secondary particles are produced under more humid conditions.

There are also processes other than those already listed which affect particle physical properties, most importantly size. The most important of them are coagulation, which results from Brownian motion and collision of particles, mainly of similar sizes, deposition of smaller particles on the surface of bigger particles, changes to the particle size owing to changes in the moisture content, including hygroscopic growth or shrinking by evaporation [6, 7]. However, typically the significance of these processes in affecting indoor particle characteristics is lower than that of processes such as deposition or removal by ventilation. This is because of the relatively long time scale of these processes and also of the relatively low particle concentration levels. For example, the process of coagulation is strongly dependent on particle concentration and while the time needed for the number concentration to halve is 0.2 s, for particle concentrations of 10^{10} cm^{-3}, which could be encountered in concentred exhaust emissions, it is as much as 55 h for concentrations of 10^4 cm^{-3}, which could be encountered in indoor environments [7].

It is very difficult to untangle the role of individual processes on particle characteristics because many of the processes take place simultaneously, affect differently particles of different size ranges and are dependent on a large number of factors and characteristics of the indoor environments. Therefore, rather than investigating the role of individual processes, many studies have considered the combined impact of several such processes on particle characteristics. An example of this is the particle loss rate, which includes surface deposition of smaller particles, due to diffusion, gravitational settling, and convective transport as well as removal of particles by ventilation.

A model developed in Ref. [29] describes particle behaviour in an indoor environment and incorporates the role of various processes. The model assumes that perfect, instantaneous mixing takes place, and also steady-state conditions in terms of outdoor concentrations and indoor source emission rates. The main model equation is

$$C_{in} = \frac{PaC_{out} + V^{-1}\sum_{i=1}^{n} Q_{is}^i}{a + k}, \tag{2}$$

where C_{in} is the indoor concentration (number of particles or mass per cubic metre of air), P is the penetration factor (coefficient), a is the air-exchange rate (per hour), C_{out} is the outdoor concentration (number of particles or mass per cubic metre of air), Q_{is}^i is the mass flux generated by the source i, n is the number of indoor sources investigated, V is the volume of the room or the house (cubic metres) and k is decay rate due to diffusion and sedimentation.

A situation often investigated is when no particles are generated from indoor sources for suitably long periods and the only contribution is from outdoor air. The indoor-to-outdoor concentration ratio of the particles is then equivalent to the infiltration factor (F_{INF}), which is defined as the equilibrium fraction of ambient particles that penetrates indoors and remains suspended [30]:

$$F_{INF} = \frac{C_{in}}{C_{out}} = \frac{Pa}{a + k}. \tag{3}$$

4
Selected Types of Indoor Pollutants

4.1
Dust

According to the terminology from aerosol science, dust is defined as solid particles formed by crushing or other mechanical breakage of a parent material, larger than about 0.5 µm [6]. According to the United States Environmental Protection Agency [31] house dust is defined as "a complex mixture of biologically derived material (animal dander, fungal spores, etc.), particulate matter deposited from the indoor aerosol and soil particles brought in by foot traffic". German Guideline 4300-8 [32] states that "there is currently no generally binding definition of the term settled house dust. To delimit the term from suspended particulate matter, it is intended to mean all types of particles, which are encountered indoors in deposited form. The dust may be solids of the most varied inorganic or organic materials which can be of natural or synthetic origin. The term includes not only fractions which originate indoors themselves, but also those which are introduced from the outside". The same document makes a distinction between old dust, which is dust of unknown age found on indoor surfaces, and fresh dust, whose age is known, and which is usually of the order of 1–2 weeks.

As already described, as a result of various human activities settled dust can be resuspended in the air. Physical and chemical properties of dust vary and depend on the source from which the dust originated.

The size distribution of house dust particles ranges over several orders of magnitude and includes particles of the order of micrometres in aerodynamic diameter as well as those of the order of several millimetres. The majority of house dust particles are of the order of tens of micrometres. For example, in Ref. [33] it was shown that 58% of dust was in the size range from 44 to 149 µm. Seifert [34] reported that 6–35% of the dust was in the range from 30 to 63 µm and in Ref. [35] it was shown that 12.8–76.4% of dust was in the range from 63 µm to 2 mm and 9.5–35.5% was smaller than 63 µm. Molhave et al. [36] reported that 41% of office dust was in the range from 50 to 125 µm and 40% was larger than 125 µm.

Dust particles of different size ranges also differ in composition and content of organic and inorganic material. The differences are related to the origin of the particles, and in particular the smaller dust particles include skin flakes, fragments of hair, microorganisms, such as fungal spores and pollen, food crumbs, abrasion of textiles and fittings, sand, loam, clay, and soot. Larger dust components can include parts of plant (e.g. leaves and needles), hair or gravel.

As can be seen from this listing, house dust contains many compounds of biological origin. The presence and the concentration of certain compounds can be used as an indicator or marker of certain biological contaminants or indoor practices. For example, the concentrations of β-$(1{\to}3)$-glucans and extracellular polysaccharides are good markers for the overall levels of fungal concentrations in floor dust according to Chew et al. [37], while endotoxin concentrations in settled dust can be used as an indicator for residential hygiene according to Bischof et al. [38].

Owing to the different origin of dust, samples collected from different indoor environments vary significantly in composition. For example, the dust from kindergartens most commonly consists almost completely of inorganic materials such as sand loam and clay from sand pits, while house dust from the residences of animal owners having at the same time heavy abrasion of carpets can consist virtually solely of organic material [4]. By contrast, dust collected in offices contains the following components: microorganisms, endotoxins, allergens, minerals and adsorbed organic compounds [36]. Residential and office cleaning removes a certain fraction of dust; however, excessive cleaning may increase the concentration of hazardous components of cleaning agents in house dust [39, 40].

4.2
Fibres

In general, fibres are particles of an elongated shape, with one dimension significantly larger than the other two. The dimension of the fibres can cover a wide range, with diameters as small as 0.025 µm and lengths reaching several hundred micrometres [41]. Fibre dust particles have been defined as those which [42]

- Exceed a length-to-diameter ratio of 3:1.
- Have a length of more than 5 μm.
- Have a diameter of less than 3 μm.

Mineral fibres are of natural and synthetic (man-made mineral fibres) origin. These two classifications, in turn, can be subdivided into inorganic and organic fibres. From the wide range of available fibres, those which are significant from both technological and health aspects are asbestos, mineral wool and ceramic fibres.

Asbestos is a collective term for silicate minerals of serpentine and amphibole groups that occur naturally as fibres. The serpentine group includes chrysotile (white asbestos, Fig. 3) and the amphibole group includes crocidolite (blue asbestos), amosite, anthophyllite, tremolite and actoinolite. Mineral wool is a term used collectively for products that consist of synthetically manufactured inorganic fibres. These vitreous (amorphic) fibres are produced from a melt and depending on the starting material are referred to as glass, rock or slug fibres. Mineral wool generally serves as heat and sound insulation in construction engineering. Ceramic fibres can have both vitreous and crystalline structures. The former are aluminium silicate fibres (refractory ceramic fibres), which can be imparted with increased temperature by the addition of certain substances (e.g. zirconium oxide) and are used, for example, in refractory linings in industrial furnaces. Crystalline ceramic fibres include aluminium fibres and single crystalline whisker fibres, which are manufactured for special applications.

The sources of asbestos and other mineral fibres in indoor environments are fire-retardants, acoustic, thermal or electric insulation and structural material. Some of the mineral fibres such as asbestos were used commonly in the past

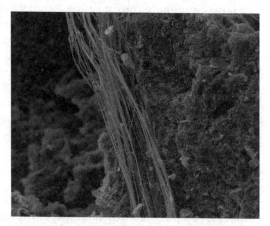

Fig. 3 Asbestos fibres (chrysotile) embedded in a magnesia plaster floor (courtesy Walter Lang, Institut für Rasterelektronenmikroskopie, Nuernberg)

as part of building construction or for insulation purposes; however, they have been banned from such applications. Exposure to fibres has been associated with serious health effects, including asbestosis, lung cancer in connection with asbestosis and mesothelioma (form of cancer).

4.3
Combustion Products in Indoor Environments

Indoor combustion sources are related mainly to cooking, heating and tobacco smoking. In addition, outdoor combustion products, which in urban environments originate most commonly from vehicle emissions, penetrate inside and contribute to indoor pollution. Under ideal conditions, complete combustion of carbon results only in the generation of CO_2 and water vapour. Any products other than CO_2 are often called products of incomplete combustion and include particulate matter and gases.

The majority of particles resulting from combustion processes are in terms of number in the ultrafine size range and in terms of mass in the submicrometre range. For example, natural gas, propane and candle flames generate particles in the size range between 0.01 and 0.1 μm and meat charbroiling shows a major peak in the particle mass distribution at 0.1–0.2 μm, with some material present at a larger particle size but not larger than 1.0 μm [16]. Aerosol particles generated through combustion can grow hygroscopically by 10–120%, depending on the initial particle size and the origin [43].

Combustion particles are of complex chemistry, carrying most of the trace elements, toxins or carcinogens generated from the combustion process. Combustion of different types of fuels results in emissions of various trace elements which are present in the fuel material. In most cases there is not just one specific element that is related to the combustion of a particular fuel, but a source profile of elements [2]. For example, motor vehicle emissions contain Br, Ba, Zn, Fe and Pb (in countries where leaded petrol is used) and coal combustion results in the emission of Se, As, Cr, Co, Cu and Al. For comparison, the crustal elements include Mg, Ca, Al, K, Sc, Fe and Mn. Since most of the trace elements are nonvolatile, associated with ultrafine particles and less prone to chemical transformations, they often remain in the air for prolonged periods of time in the form in which they were emitted.

All of the combustion sources generate large amounts of volatile and semivolatile organic compounds. Polynuclear aromatic hydrocarbons (PAH), some of which are strongly carcinogenic, are an important class of compounds contained in the organic fraction of fine particulate matter. PAH compounds are synthesised from carbon fragments into large molecular structures in low-oxygen environments, such as occurs inside the flame envelope in the fuel-rich region of the flame structure [2]. If the temperature is not adequate to decompose compounds upon exiting from the flame zone, then they are released into the free atmosphere and condense or are adsorbed onto the surface of particles. Many different combustion systems are known to produce

PAH compounds. The most studied PAH is benzo[a]pyrene, which is a physiologically active substance that can contribute to the development of cancer in human cells.

Semivolatile organic compounds can be present in the air either in vapour or in particle form (solid or liquid). From the point of view of the effect on human health it could be of significance in what physical form the semivolatile compounds are when they are inhaled. There is very little information available on this aspect, which is due not only to the recency of the interest, but mainly to difficulties in investigating the organic composition of small amounts of material. The mass of the particles in the submicrometre and ultrafine range is very small, and in order to collect sufficient mass for standard organic chemistry analyses, long sampling times are required, which is prohibitive for many exposure or health effects studies.

4.3.1
Environmental Tobacco Smoke

Almost all of the major studies have found that an important source of fine particles is cigarette smoking, resulting in an increase in the average fine particle levels indoors in the range from 10 to 45 µg m^{-3} [44]. Thus environmental tobacco smoke (ETS) is one of the most significant indoor pollutants. ETS is a mixture of two components: mainstream smoke drawn through the tobacco, taken in and exhaled by the smoker, and sidestream smoke, which is emitted by the smouldering cigarette between puffs. Exposure by nonsmokers – passive smoking – is of great concern as it is an involuntary risk, often incurred by the most susceptible members of society, unborn children, infants and young children. Over 4,000 compounds have been identified in ETS, the most important of them being CO, NO_x, nicotine, acetone, benzene, phenol, toluene, formaldehyde and benzo[a]pyrene [16]. Other organic compounds present in ETS include isoalkanes, anteisoalkanes (anteisotrioacontane, anteisohentriacontane, anteisodotriacontane and isotritracontane) [45]. Once generated and introduced into the air, both gas and particle phases interact with each other (gas-to-particle conversion and particle–particle interactions), with atmospheric aerosols and with the environment.

4.3.1.1
Physical Characterisation of the Particulate Phase of Cigarette Smoke

Particles generated from combustion of a cigarette range in size quite substantially, with the vast majority of them, however, being very small, below 1 µm, and a significant fraction below 0.1 µm. A microscope image of an ETS particle is presented in Fig. 4. A literature review conducted recently [44] revealed that most studies reported ETS particles as occurring in a single size mode, with a large majority of the particle mass distributed among particles with diameters in the range 0.02–2 µm. The median of the MMDs reported in the studies review-

Fig. 4 Environmental tobacco smoke (*ETS*) particle examined with an energy-dispersive X-ray analyser attached to a transmission electron microscope. The dominant element of the particle is carbon

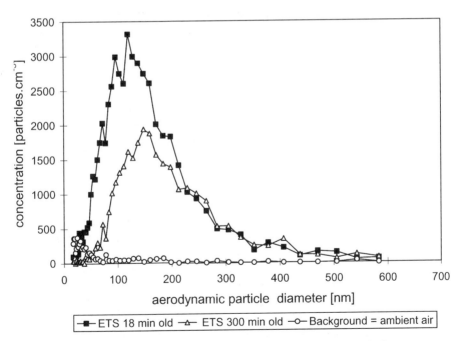

Fig. 5 Size distribution of ETS measured 18 and 300 min after generation [53]

ed was found to be 0.37 µm, with the central 90% ranging between 0.2 and 0.5 µm. The median reported geometric standard deviation was 1.4, with the central 90% ranging between 1.2 and 2.1 µm. An example of a number size distribution of ETS measured in an experimental chamber is presented in Fig. 5 [53].

The analysis conducted by Nazaroff and Klepeis [44] also showed that the median emission factor for cigarette smoking of the reported data on ETS is 12.7 mg per cigarette, with the distribution exhibiting positive skew. The lognormal parameters of the distribution were a geometric mean of 12.0 mg per cigarette, with a geometric standard deviation of 2.1. The central 90% of the distribution lay between approximately 5 and 40 mg per cigarette, with ten of the 17 results being in the range 7.8–13.8 mg per cigarette. The total mass of particulate matter emitted from cigarettes in a particular indoor environment can be estimated as the product of such emission factors and the number of cigarettes smoked indoors.

4.3.1.2
Chemical Characterisation of Cigarette Smoke

In addition to inorganic gases such as carbon monoxide and nitrogen oxides a large number of elements and compounds are generated from cigarette combustion. The particulate phase of the smoke is of special importance because it contains a significant amount of cigarette combustions products, for example, all of the tar and most of the nicotine [46]. Some of the compounds are present in cigarette smoke in very minute quantities, often difficult to measure; some of them, however, are emitted at much higher concentrations. Those emitted at higher concentrations and which are specific to cigarette smoke are called markers of the smoke. Nicotine, carbon monoxide, 3-ethenylpyridine (3-EP), nitrogen oxides, pyridine, aldehydes, acrolein, benzene, toluene and several other compounds have been used or have been suggested for use as markers for vapour-phase constituents of cigarette smoke. Repairable suspended particulate matter (the mass fraction of inhaled particles which penetrates into the unciliated airways), solanesol, N-nitrosamines, cotinine, chromium and potassium are among the air contaminants used as markers for the particle-phase constituents of the smoke [47].

The substances most commonly utilised as markers of ETS are repairable suspended particulate matter, nicotine, CO, 3-Ethenylpyridine (3-EP) and solanesol [48, 49]. All these substances are associated, however, with potential problems, when used as markers. Both CO and repairable suspended particulate matter are not unique to ETS. The use of nicotine as a marker of ETS presents a problem because [50] (1) nicotine is found primarily in the gas phase (90%), making it a relatively poor particle marker, (2) gas-phase nicotine is strongly basic and is removed from indoor environments at a faster rate than particle-phase nicotine or the particle portion of ETS and (3) the fraction of nicotine in ETS varies with measurement conditions. For example, 5–10% of ETS nicotine was found in the particle phase in a controlled atmosphere, while 20% was found in field environments. 3-EP and solanesol are currently considered the best available ETS markers for the vapour and particulate phases, respectively [49]. The problems, however, are that airborne concentrations of 3-EP do not increase with source strength (i.e. with the number of cigarettes smoked) and that the methods for 3-EP determination do not possess adequate sensitivity [49].

In addition to the limitations already discussed in using various substances as ETS markers, consideration should also be given to the degree of complexity associated with using a particular substance as a marker and also whether it could be measured in real time (to enable immediate mitigation actions if necessary). Respirable suspended particulate matter, CO and NO_x are the only markers for which measurements can be performed on a real-time basis with existing commercial equipment. However, as already explained, none of them are specific to ETS. Often the majority of CO and NO_x comes from sources other than ETS [51]. The utilisation of solanesol presents a limitation because it cannot be measured in real time and what is more, its determination is not simple. It involves extraction from a filter to a solution with a recovery of about 60–90% and sample analysis by gas chromatography and supercritical fluid chromatography [52].

In summary, the chemistry of a cigarette smoke is very complex, with a vast number of elements and compounds generated in particle, vapour and gaseous phases. Some of these elements and compounds are generated in very small quantities, while others are in larger, easily measurable quantities. Some of the combustion products are specific to tobacco smoking, such as nicotine, others are also emitted by different, particularly combustion, sources. These compounds which are specific to ETS have been used as a marker of the smoke, but also compounds that are nonspecific to the smoke have been used for this purpose as well. The applicability of a compound as a marker of cigarette smoke does not necessarily point out the health impact of this particular compound.

4.3.1.3
Changes Occurring to Cigarette Smoke After Generation

After generation, the highly dynamic and reactive mixture of combustion products undergoes substantial changes resulting from complex physical and chemical processes taking place. The most substantial changes occur immediately after generation when the concentration of combustion products is still high. With time the concentration decreases owing to the dilution with atmospheric air and the rate of most of the processes decreases. Removal of the combustion products from the air is mainly due to ventilation and filtration processes, with other processes, for example, surface deposition playing a lesser role. This applies both to the particle and gaseous phases. The particles generated from cigarette combustion change their size distribution mainly due to coagulation with each other, which shifts the peak of the distribution towards larger sizes. They can remain suspended in the air for long periods of time, up to several hours, as the process of gravitational deposition is not very efficient in this particle size range.

Variations in peak shape and location between the measurements taken at 10 and 180 min after the introduction of cigarette smoke to an experimental chamber are presented in Fig. 6 [53]. It can be seen that during the latter measurement, the ETS peak is still clearly distinguishable from the background environmental aerosol spectrum and its location is shifted towards a larger size compared with the initial location.

Changes occurring in ETS in a real indoor environment were investigated by Morawska et al. [53], who measured the ETS decay in residential houses for conditions of minimum and normal ventilation. For minimum ventilation (all the windows and doors closed) the air-exchange rate was estimated to be in the range 0.55–0.79 h^{-1} for brick houses and 1.05 for a wooden house, and for normal ventilation (defined as all the windows which are normally open), from 1.93 to 4.48 h^{-1} and 4.73 h^{-1} for brick and wooden houses, respectively. The dependence of the particle CMD and concentration on time for minimum ventilation in a brick house is presented in Fig. 5. Inspection of Fig. 5 reveals that for minimum ventilation in a brick house a distinctive ETS peak is present in the air, even 3 h after one cigarette had been smoked. Measurements conducted in a wooden house (higher air-exchange range) showed that the peak is present in the air for up to 2 h [43] and drew similar conclusions as to the concentration changes in the submicrometre ETS peak in a residential location of low air-exchange rate. For normal ventilation the ETS peak was distinguishable for short periods of time, which was about 1 h for a brick house and less than half an hour for the wooden house. When the concentration remained high for a longer time (as in the brick house at minimum ventilation) the CMD of the peak increased from about 90 nm to about 150 nm. At lower concen-

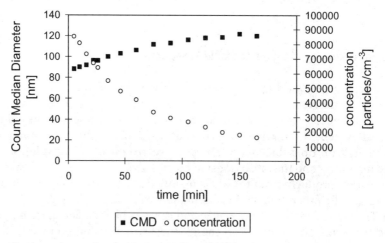

Fig. 6 The dependence of particle count median diameter (*CMD*) and concentration on time for minimum ventilation in a brick house [53]

trations, the size distribution did not markedly change, which can be explained by a slowing down of the coagulation process and particle growth at lower concentrations.

Measurements of ETS characteristics in a large, naturally ventilated university club showed that at most times when the number of smokers was small (less than ten), the smoke concentration decreased rapidly with time and the ETS peak was not detectable even 10–15 min after a cigarette had been smoked [53]. However, when the number of smokers was large, for example, during a rock concert, the ETS concentration in the room was very high, up to 5×10^4 particles cm^{-3} and was characterised by the stable, well-defined broad peak.

4.3.2
Biomass Burning

Biomass burning in indoor environments is most commonly conducted for the purpose of cooking or heating, using as fuel wood and in developing countries also animal dung, crop residues or charcoal. Household cooking stoves and space heaters are counted in billions throughout the world, providing the very basic household needs of heat; however, they are a source of significant levels of pollutants, when dirty biomass fuels and inadequate technologies are used. The levels of particles, PAHs, CO, and other air pollutants found in the kitchens of developing-country villages are orders of magnitude above Western urban levels or relevant standards [3, 54, 55]. In general, because the household sources emit directly into the indoor spaces and at the times of human occupancy, exposures to certain air pollutants derived from these small sources are often greater than those derived from large outdoor sources [56–58].

The majority of particles emitted from biomass burning are ultrafine, with only a small fraction in the larger size range, and with most of the mass present in the particles less than 2.5 µm in aerodynamic diameter [59]. For example, Raiyani et al. [60] investigated emissions from burning several biomass fuels in cooking stoves. The fuels tested included cattle dung, wood, crop residues, low-quality charcoals and also coal. The study revealed that 50–80% of TSP emissions from these cooking stoves were 2 µm or smaller and also that a large amount of the PAHs (above 75%) were found in this size fraction. A number of studies investigated the characteristics of emissions from residential wood burning stoves, which are common in the USA for space heating or aesthetic purposes. The studies reported that (1) the particle mass distribution from wood (pine, oak, eucalyptus) combustion has a single mode at approximately 0.1–0.2 µm [16], (2) the particles are compact structures with fractal-like dimensions close to 3 and contained low mass fractions of volatile compounds [61] and (3) that operating conditions, such as the air supply, had a strong impact on the particle size distribution and the emission of particle-bound PAHs [61].

Vegetation burning results in emission of a large number of compounds, including retene, phytosterols, ligmens, phenalic compounds from lignins and diterponoids from resins [45]. The emissions from wood-burning stoves were found to be acidic (pH 2.8–4.2) [62]. Organic acids, of which the major constituents are monocarboxylic (emitted from combustion of fossil fuels and biomass) and diacarboxylic acids [63], have been linked to health effects. High concentrations of PAHs have been found in soot generated from wood-burning stoves [64], with some of them specific to biomass burning. This makes it is possible to use certain PAHs as source signatures in receptor modelling for residential wood combustion [65]. For example, guaiacol and its derivatives (e.g. 4-methylguaiacol, 4-ethylguaiacol) result solely from the pyrolysis of wood lignin, are relatively stable in the atmosphere and therefore can serve as unique tracers of wood [66]. In terms of elemental composition, noticeable elemental carbon was found in wood smoke as well as measurable quantities of Na, K, Fe, Br, Cl, nitrate, sulfate and ammonium. Statistically significant amounts of Na, Al, K, Sr, Ba, Cl, nitrate and sulfate were found in meat-charbroiling emissions [16].

4.3.3
Vehicle Emissions

As already discussed, outdoor air has a significant impact on indoor air, with up to 100% penetration efficiency of pollutants from outdoor air to indoor air for naturally ventilated buildings. For mechanically ventilated buildings, the fraction of pollutants penetrating indoors is smaller and depends on a number of factors, including the type and operation of the filtration and ventilation system, the type of filters used, and thus their efficiency for particles in different size ranges, the location of the air intake, and tightness of the building. Owing to the significant effect of motor vehicle emissions on outdoor air, and efficient penetration of outdoor pollutants into indoor environments, the impact of vehicle emissions on indoor concentration levels of particles and other pollutants could be substantial.

Vehicle emissions, like other combustion products, are composed of pollutants in gaseous and particle forms, which are complex in chemistry, and contain many compounds which have been shown to affect human health. The main gaseous emissions include hydrocarbons, CO, NO_x, CO_2, SO_2 and water vapour. Particles generated from vehicle emissions are generally small. A significant proportion of diesel emission particles have diameters smaller than 0.1 μm [67]. Gasoline particles are mostly carbonaceous spherical submicrometre agglomerates ranging from 10 to 80 nm. Particles from compressed natural gas emissions are smaller than from diesel or even petrol emissions and range from 0.01 to 0.7 μm, with the majority being between 0.020 and 0.060 μm. [68]. Associated with particles (especially fine and ultrafine) are many toxins, trace elements and also carcinogenic compounds. An example of these is 3-nitrobenzanthrone, a nitrated PAH originating from diesel emissions which has been shown to have high cancer-causing potential.

The chemistry of particles originating from vehicle emissions varies and depends on engine technology, test conditions, type of fuel on which the vehicle operates, its specific composition and on other characteristics, as well as the lubricating oil used and its composition. There are thus differences between particles originating from diesel or spark ignition vehicles, the latter including petrol, compressed natural gas, liquid petroleum gas or, for example, ethanol-fuelled vehicles. Diesel emission particles are primarily elemental carbon, but they also contain adsorbed or condensed hydrocarbons, hydrocarbon derivatives, sulfur compounds and other materials [69]. Solvent extractable organic components of diesel aerosols represent 5–40% of the particle mass. Particles emitted from spark ignition vehicles are mostly carbonaceous spherical sub-micrometre agglomerates, consisting of a carbon core with various associated organic compounds. The main components of the particle phase include soot and ash, which consist of trace elements such as lead, iron, chlorine and bromine, organic compounds and a low-to-medium boiling fraction of engine oil [70]. Lubricating oil and other fuel hydrocarbons are the main contributors to emissions of particles of nanometre size [71]. The sulfate particles present in gasoline engine emissions are mainly from catalyst-equipped vehicles utilising unleaded gasoline [72].

In general, motor vehicle emissions contain various organic compounds, including hopanes and steranes (present in lubricating oil for diesel and gasoline vehicles, and in diesel) and black elemental carbon (present in a higher fraction in diesel emissions) [45]. Semivolatile aliphatic hydrocarbons present in emissions from diesel and gasoline engines consist of a narrow band of C_{15-27} n-alkanes maximising at C_{20-21}, a very similar pattern to lubricating oils n-C_{13-27}, maximising at C_{19}. The signal of diesel fuel has a broader spectrum extending to n-C_{33}, with a higher proportion of lower molecular weight components (n-C_{10-22}, maximising at C_{19}) [73]. Common organic compounds are PAHs, such as pyrene, chrysene and benzo[a]pyrene. The semivolatile fraction of the emissions can be associated either with vapour or with particle phases.

References

1. Wallace L (2000) Aerosol Sci Technol 32:15
2. Morawska L, Zhang J (2002) Chemosphere 49:1045
3. Zhang J, Morawska L (2002) Chemosphere 49:1059
4. Morawska L, Salthammer T (2003) In: Morawska L, Salthammer T (eds) Indoor environment, airborne particles and settled dust. Wiley-VCH, Weinheim, pp
5. Özkaynak H, Xue J, Weker R, Butler D, Koutrakis P, Spengler J (1994) The particle TEAM (PTEAM) study: analysis of the data – final report, vol III. Harvard School of Public Health, Boston, MA
6. Willeke K, Baron PA (1993) In: al e (ed) Van Nostrand Reinhold, New York, pp
7. Hinds WC (1982) Aerosol technology: properties, behavior, and measurement of airborne particles. Wiley, New York
8. Whitby KT (1987) Atmos Environ 12:135
9. Schmidt-Ott A (1988) Appl Phys Lett 52:954

10. Peitgen HO, Richter PH (1986) The beauty of fractals. Springer, Berlin Heidelberg New York
11. Wolfram S (2002) A new kind of science. Wolfram Media, Champaign
12. Ramsak P, Veld PO (2000) In: al e (ed) Healthy buildings 2000: design and operation of HVAC systems, vol 2. SIY Indoor Air Information, Helsinki, p 647
13. Abt E, Suh HH, Allen G, Koutrakis P (2000) Environ Health Perspect 108:35
14. Li C-S, Jenq F-T, Lin W-H (1992) J Aerosol Sci 23:s547
15. Li C-S, Lin W-H, Jenq F-T (1993) Atmos Environ 27B:413
16. Kleeman MJ, Schauer JJ, Cass GR (1999) Environ Sci Technol 33:356
17. Morawska L, Thomas S, Bofinger ND, Wainwright D, Neale D (1998) Atmos Environ 32:2461
18. Morawska L (2000) In: Healthy buildings 2000, Espoo, Finland
19. Chao C, Tung T, Burnett J (1998) Indoor Built Environ 7:110
20. Matthias-Maser S, Jaenicke RT (1995) Atmos Res 39:279
21. Wallace L (1996) J Air Waste Manage Assoc 46:98
22. Mosley RB, Greenwell DJ, Sparks LE, Guo Z, Tucker WG, Fortmann RC, Whitfield C (2001) Aerosol Sci Technol 34:127
23. Long CM, Suh HH, Catalano PJ, Koutrakis P (2001) Environ Sci Technol 35:2089
24. Thatcher TL, Layton DW (1995) Atmos Environ 29:1487
25. Yamamoto T, Ensor D, Sparks L (1994) Build Environ 29:291
26. Jamriska M, Morawska L, Ensor D (2003) Indoor air (in press)
27. Weschler CJ, Shields HC (1999) Atmosp Environ 33:2301
28. Weschler C (2003) In: Morawska L, Salthammer T (eds) Indoor environment, qirborne particles and settled dust. Wiley-VCH, Weinheim, pp
29. Koutrakis P, Briggs SK, Leaderer BP (1992) Environ Sci Technol 26:521
30. Wilson WE, Mage DT, Grant LD (2000) J Air Waste Manage Assoc 50:1167
31. USEPA (1997) Exposure factors handbook. National Center for Environmental Assessment, Washington, DC
32. Verein Deutscher Ingenieure (2001) VDI 4300-8. Beuth, Berlin
33. Quee Hee S, Peace B, Clark CS, Boyle JR, Boule JR, Bornschein RL, Hammond PB (1985) Environ Res 38:77
34. Seifert B (1998) Bundesgesundheitsblatt 41:383
35. Salthammer T (2003)
36. Molhave L, Schneider T, Kjaergaard SK, Larsen L, Norn S, Jorgensen O (2000) Atmos Environ 34:4767
37. Chew GL, Douwes J, Doekes G, Higgins KM, van Strien R, Spithoven J, Brunekreef B (2001) Indoor Air 11:171
38. Bischof W, Koch A, Gehring U, Fahlbusch B, Wichmann HE, Heinrich J (2002) Indoor Air 12:2
39. Wolkoff P, Schneider T, Kildesø J, Degerth R, Jaroszewski M, Schunk H (1998) Sci Total Environ 215:135
40. Vejrup K, Wolkoff P (2002) SciTotal Environ 300:51
41. International Labour Organisation (1997) International Labour Organisation, Geneva
42. Bake D (2003) In: Morawska L, Salthammer T (eds) Indoor environment, airborne particles and settled dust. Wiley-VCH, Weinheim, pp
43. Li W, Hopke PK (1993) Aerosol Sci Technol 19:305
44. Nazaroff W, Klepeis N (2003) In: Morawska L, Salthammer T (eds) Indoor environment, airborne particles and settled dust. Wiley-VCH, Weinheim, pp
45. Cass GR (1998) Trends Anal Chem 17:356

46. First MW (1985) In: Gammage RB, Kaye SV (eds) Indoor air and human health. Lewis, MI, pp
47. Leaderer BP, Hammond SK (1991) Environ Sci Technol 25:770
48. Rando RJ, Menon PK, Poovey HG, Lehrer SB (1992) Am Ind Hyg Assoc J 32:3845
49. Odgen M, Heaven D, Foster T, Maiolo K, Cahs S, Richardson J, Martin P, Simmons P, Conrad F, Nelson P (1996) Environ Technol 17:239
50. Eatough DJ, Benner CK, Tang H, Landon V, Richards G, Caka FM, Crawford J, Lewis EA, Haasen LD, Eatough NL (1989) Environ Int 15:19
51. Eatough DJ, Hansen LD, Lewis EA (1988) Indoor and ambient air quality. Sepler, London, p 131
52. Tang H, Richards G, Benner C (1990) Environ Sci Technol 24:848
53. Morawska L, Jamriska M, Boginger ND (1997) Sci Total Environ 196:43
54. Zhang J, Smith KR (1996) J Exposure Anal Environ Epidemiol 6:147
55. Smith KR (1986) Environ Manage 10
56. Saksena S, Prasad R, Pal RC, Joshi V (1992) Atmos Environ 26A:2125
57. Smith KR (1993) Annu Rev Energy Environ 18:529
58. Zhang J, Smith KR (1999) Environ Sci Technol 33:2311
59. World Health Organization (1999) World Health Organization health guidelines for vegetation fire events. World Health Organization, Geneva
60. Raiyani CV, Shah SH, Desai NM, Kenkaiah K, Patel JS, Parikh DJ, Kashyap SK (1993) Atmos Environ 27A:1643
61. Hueglin C, Gaegauf C, Kunzel S, Burtscher H (1997) Environ Sci Technol 31:3439
62. Burnet PG, Edmisten NG, Tiegs PE, Houck JE, Yoder RA (1986) Air Pollut Control Assoc 1012
63. Limbeck A, Puxbaum H (1999) Atmos Environ 33:1847
64. Mumford JL, Harris DB, Williams K, Chuang JC, Cooke M (1987) Environ Sci Technol 21:308
65. Li CK, Kamens RM (1993) Atmos Environ 27A:523
66. Hawthorne SB, Miller D, Langenfeld JJ, Keieger MS (1992) Environ Sci Technol 26:2251
67. Morawska L, Bofinger N, Kosic L, Nwankowala A (1998) Environ Sci Technol 32:2033
68. Ristovski Z, Morawska L, Thomas S, Hitchins J, Greenaway C, Gilbert D (2000) J Aerosol Sci 31:403
69. Kittelson DB (1998) J Aerosol Sci 29:525
70. Zinbo M, Korniski TJ, Weir JE (1995) Ind Eng Chem Res 34:619
71. Kittelson DB, Watts WF, Johnson JH (2002)
72. Brodowicz P, Carrey P, Cook R, Somers J (1993) In: EPA Technical Support Branch, Emission Planning and Strategies Division, Office of Mobile Sources, Ann Arbor, MI, pp
73. Simoneit BRT (1985) Int J Anal Chem 22:203

The Handbook of Environmental Chemistry Vol. 4, Part F (2004): 149–177
DOI 10.1007/b94834
© Springer-Verlag Berlin Heidelberg 2004

Indoor Air Pollution by Microorganisms and Their Metabolites

Hans Schleibinger (✉)[1] · Reinhard Keller[2] · Henning Rüden[1]

[1] Institute of Hygiene and Environmental Medicine, Charité, University Medicine Berlin, Hindenburgdamm 27, 12203 Berlin, Germany
hans.schleibinger@gmx.de
[2] Institute of Hygiene, University Hospital Schleswig-Holstein, Campus Lübeck, Ratzeburger Allee 160, 23538 Lübeck, Germany

Abstract Microbial damage in indoor areas is caused most frequently by molds and bacteria. These microorganisms have a very important role in the biogeochemical cycle, as their task consists in disintegrating organic mass to reusable metabolites. In the environment spores of molds and bacteria may become airborne and are therefore ubiquitous. They can enter indoor areas by passive ventilation or by ventilation systems as well. Many genera are also emitted by indoor sources like animals, flowerpots and wastebaskets. This normal flora is in most cases not harmful. But more and more frequently growth conditions like excessive humidity and/or a high water content of building materials are encountered, which are in most cases the limiting factor for microbial growth. This is caused by shortcomings of the buildings like the lack of thermal insulation as well as by the noncorrect behavior of the room users. According to the relative humidity and/or the moisture content of the building material different microorganisms are able to grow on indoor building materials and may cause destruction, adverse health effects and unpleasant smells. Therefore the task of microbial examinations is to differentiate between normal indoor microorganisms, airborne or adherent to walls and floors, and between more or less heavily growing species, attacking building materials and producing microbial products causing adverse health effects. Air sampling of microorganisms is a very favored method, as it allows a direct toxicological evaluation, as results can be related to a concentration expressed in colony forming units per cubic meter. Sometimes even information on the particle size is available, which allows an estimation of how deep those particles may penetrate into the lung. But microorganisms in indoor air are generally not equally distributed, but occur in clouds, so they are often overlooked in air measurements, especially if the microbial damage is hidden by paneling, walls, etc. Another reason for false-negative results obtained by air measurements is that fungal spores are not released at all stages of growth. Therefore other techniques are helpful for example, the sampling of household dust, the sedimentation method or direct sampling from surfaces. The differentiation of bacteria is performed by biochemical methods as a rule, whereas in most cases the differentiation of molds is done by microscopy, especially by the forms of spores. On many occasions the growth behavior and patterns on different nutrient agars also have to be evaluated. Nonsporulating species have to be triggered to produce spores, otherwise "sterile mycelium" will be indicated, which means they cannot be named by genera or even species. Methods of genetic fingerprinting are still in their early stages and only available for some genera or species. But in the meantime enzymatic tests have become available to decide between mold growth and normal quantities on building surfaces. A very difficult task consists in the search for hidden mold growth, for example, if adverse health effects like the fungal syndrome are observed, which is characterized by the occurrence of unspecific symptoms. There are efforts to detect hidden mold growth by the analysis of microbial volatile organic compounds or by the use of specially trained sniffer dogs, but these methods have not been scientifically evaluated. The odor alone perceived by human beings is not reliable to detect mold damage. As far as the rehabilitation of the indoor environment is concerned, it has to be pointed out very clearly that microbial damage has to be removed. The killing of microorganisms is often carried out, but this procedure is not sufficient, as for example, nonviable spores keep their allergenic potential. The acuteness of the rehabilitation procedures is normally considered according to the extent of the microbial damage. Adverse health effects are supposed to be connected with microbial growth in indoor areas, mostly with mold growth. Predominantly allergies have to be mentioned, followed by toxic alveolitis and reactions like (allergic) bronchitis, chronic obstructive pulmonary disease, but also the aggravation of asthma. Infections by molds and bacteria are very rare, but persons with an immunodeficiency are especially susceptible to fungal infections. It has been found that spores of fungi contain fungal toxins (mycotoxins), which are well known from food contaminations. But whether these mycotoxins show toxic effects, if fungal spores are inhaled, has not been confirmed. On the whole, the dose relationship between the concentration of microbial particles mentioned and the adverse health effects described is not very well established. When sanitary effects are observed, very often the susceptibility of the individual is crucial. Therefore guidelines concerning microbial products in indoor areas are sparse and mostly not scientifically sound.

Keywords   Microbial damages in indoor environment · Molds · Bacteria · Measurement strategy · Health effects

1
Introduction: Natural Occurrence of Bacteria and Molds

1.1
Function in the Ecosystem

Bacteria and molds are highly diverse and versatile microorganisms that have adapted to almost all environments. They are ubiquitous and occur in and on animals, plants and humans as parasites or symbionts. In nature, prokaryotic bacteria and archaea as well as eukaryotic molds play a key role in the biogeochemical cycle. They are saprogenic organisms involved in the recycling of organic matter. Thus, they play an important role in the stabilization of the balance between organic and inorganic compounds.

Especially in soil, fungi have the task of destroying cellulose, a structural substance of plants and one of the most abundant organic compounds on earth, and of making the decomposition products available to other microorganisms. Both fungi and bacteria colonize all substrates meeting their physiological growth and reproduction demands. Under certain conditions this may lead to the contamination and destruction of "man-made" environments, with unwanted microorganisms growing in private dwellings, work environments and production processes. Microbial growth in indoor environments may cause adverse health effects such as allergies and asthma (see Sect. 5).

1.2
Nomenclature

As to the nomenclature, the first name of a microorganism describes the affiliation to a genus (e.g., *Aspergillus*, *Penicillium* or *Bacillus*) and the second one the species (e.g., *Aspergillus fumigatus*, *Penicillium brevicompactum* or *Bacillus subtilis*). It is scientifically correct and very helpful to differentiate between the species by stating the full name, for example, when comparing different studies. However, many laboratories only indicate the genus, especially with molds, since differentiating down to the species is often time-consuming and expensive (e.g., transferring the colony under question to different petri dishes), requires great skill and is often not paid for by the contractor. Differentiation of the genus *Penicillium* is especially difficult because many species are very similar. It is generally sufficient to know the spectrum of the genera for simple reports such as those on mold damage. However, differentiation of the species is essential and indispensable for addressing health complaints or when expert reports or legal advice are necessary. The genera and species of molds and bacteria are generally printed in italic letters. With well-known species, the first name (the genus) is often abbreviated, especially, if the genus was mentioned already, for example, *A. fumigatus* instead of *Aspergillus fumigatus*.

1.3
Prevalence of Genera in Outdoor and Indoor Air

Spores of microscopic fungi, commonly called molds, and bacteria are always present in the air. There is a wide range in the number and type of molds and bacteria found in outdoor air depending on factors like the time of day, weather, season, geographical location and occurrence of spore sources. According to Ref. [1] *Cladosporium* and *Alternaria* are the most abundant molds in outdoor air in temperate climates. Mold spores and bacteria are ventilated into indoor air from the outside and start to grow on walls, carpets, furniture, all types of organic material, house dust or food, if the appropriate conditions are encountered (see Sect. 2.1). Spore concentrations of molds like *Cladosporium*, *Alternaria*, *Epicoccum* and *Botrytis* encountered in indoor air differ from spore concentrations outside the building [2, 3]. This means that the amount and the spectrum of molds and bacteria found in buildings are strongly influenced by people, pets, plants, ventilation systems and the season. The findings are also dependent on the type of indoor air sampling [1]. In a review [4] wide variations were reported in the amount of viable and cultivable molds and colony forming units (cfu) in the indoor air of moldy and normal households. Up to 60 species of microscopic fungi were found in indoor air. Thus, mold damage can be predicted in many cases by measuring air or examining other materials like household dust (see Sect. 3).

2
Microbial Damage in Indoor Environments

2.1
Growth Conditions in Indoor Spaces

Prerequisites for microorganism growth are nutrient sources, a certain temperature range, the pH value of the substrate, the amount of free water, the absence of inhibiting substances or competing species and sometimes light conditions. Molds generally have low demands concerning oxygen in the atmosphere.

Thus, some species of *Mucor* and *Rhizopus* can germinate in an atmosphere of pure nitrogen. Other common molds like *Alternaria* and *Cladosporium* are able to grow with only a trace of oxygen [5, 6].

Since mold spores and bacteria are often airborne and thus ubiquitous and generally require only few nutrient media and other parameters, the amount of water in indoor spaces is usually the factor limiting their growth. As carbon-heterotrophic microorganisms, molds are dependent on a supply of nutrient substrates (e.g., cellulose) that are absorbed in dissolved form from the environment. Enough water in a liquid state must be available for this [7]. The total water content of the substrate is not available but only the part that is not bound to dissolvable substances (salts, carbohydrates, proteins). The water availability is designated as water activity (a_W value) [6] and is dependent on the substrate temperature. The water activity is defined as the quotient of the water vapor pressure over the substrate (P_D) and the saturation pressure (P_S) of pure water at a given temperature in a enclosed system:

$$a_W = P_D/P_S .$$ (1)

The a_W value is also designated as the relative equilibrium moisture. The following formula is used to express the relationship between the a_W value and the relative humidity (RH), which is in equilibrium over the substrate [8]:

$$RH (\%) = a_W \times 100 .$$ (2)

The lowest a_W values at which the microorganisms can grow or germinate are listed in Table 1.

The water activity can be given for the substrate (e.g., for a specific building material or the nutrient in a petri dish) or as a threshold value for the optimal and minimal growth of microorganisms or the formation of spores. The highest a_W value of 1.0 is assigned to distilled water and absolutely dry materials like glass have an a_w value of 0.0. Examples of the water content needed for growth of selected molds on typical indoor materials [10] are presented in Table 2 together with minimal water activities of some molds [7, 11]. Molds frequently infest building materials in a certain order. Thus, there are primary, secondary and tertiary colonizers. In this way, the age of the damage can sometimes be determined from the spectrum of the genera.

Table 1 Lowest a_W value, at which microorganisms can grow or germinate [9]

Microorganism group	Lowest a_W value
Bacteria	0.91
Yeasts	0.88
Molds	0.80
Halophilic bacteria	0.75
Xerophilic molds	0.65
Osmophilic yeasts	0.60

Table 2 Minimal water activity for the growth of molds on building materials, paints, finishings and furnishings [10] and minimal water activities of molds on nutrient agar [7, 11]. RH_{EC} is the relative humidity in an equilibrium condition

Water activity (RH$_{EC}$)	Microorganisms [10] Growth on building materials, paints, finishings and furnishings	Microorganisms [7, 11] Growth on nutrients
High: $a_w > 0.90$ (RH$_{EC} > 90\%$)	**Tertiary colonists** (hydrophilic species): *Mucor plumbeus* *Alternaria alternata* *Stachybotrys chartarum* Yeasts, for example *Rhodotorula* *Sporobolomyces* Actinomycetes	*Aspergillus fumigatus* *Stachybotrys chartarum*
Moderate: $a_w = 0.80–0.90$ (RH$_{EC} = 80–90\%$)	**Secondary colonists** (mesophilic species): *Cladosporium cladosporioides* *Cladosporium sphaerospermum* *Aspergillus flavus* *Aspergillus versicolor*	*Alternaria alternata* *Alternaria citri* *Aspergillus flavus* *Aspergillus nidulans* *Aspergillus wentii* *Cladosporium herbarum* *Cladosporium cladosporioides* *Epicoccum nigrum* *Fusarium culmorum* *Mucor circinelloides* *Paecilomyces variotii* *Penicillium brevicompactum* *Penicillium chrysogenum* *Penicllium expansum* *Penicillium frequentans* *Penicillium griseofulvum* *Penicillium spinulosum* *Penicillium citrinum* *Rhizopus stolonifer* *Rhizopus microsporus* *Rhizopus oryzae* *Saccharomyces cerevisiae*

Table 2 (continued)

Water activity (RH_{EC})	Microorganisms [10] Growth on building materials, paints, finishings and furnishings	Microorganisms [7, 11] Growth on nutrients
Low: $a_w < 0.80$ ($RH_{EC} < 80\%$)	**Primary colonists** (xerophilic species): *Aspergillus versicolor* *Aspergillus glaucus* *Aspergillus penicillioides* *Penicillium brevicompactum* *Penicillium chrysogenum* *Wallemia sebi*	*Aspergillus candidus* *Aspergillus amstelodami* *Aspergillus niger* *Aspergillus ochraceus* *Aspergillus repens* *Aspergillus restrictus* *Aspergillus sydowii* *Aspergillus terreus* *Aspergillus versicolor* *Eurotium chevalieri* *Wallemia Sebi* *Emericella nidulans* *Fusarium solani*

2.2
Buildings: Structural Shortcomings, Dew Point, Lack of Thermal Insulation, Air Exchange Rate

High air humidity is not the only reason for microbial growth indoors. Microbial growth is often caused or at least promoted by moisture penetration through building materials. The causes of moisture penetration can be divided into construction flaws or user errors. Thermal bridges and defective isolation are considered as construction flaws.

2.2.1
Moisture Penetration Caused by Thermal Bridges

Thermal bridges are a major problem especially in old buildings. They are structural areas in which there is a higher heat flow than in the surroundings. Owing to e.g. defective isolation, the heat flows from warmer to colder areas. The thermal site can then cool below the dew point, where water condenses from the water vapor of the air and then is available in the form of free water for the growth of the microorganisms. Moreover, thermal bridges increase the deposition of dust, which is a nutrient source promoting microbial growth. The reasons for thermal bridges are given in Table 3. However, dilapidated heating systems, false radiator placement or reduced thermal convection can also lead to cool areas below the dew point. Frequent causes of mold growth also include water damage due to leaks from cracks in the building facade, the roof or water pipes.

Table 3 Causes of thermal bridges [12]

Thermal bridges caused by	Example/explanation
Geometry	Due to large construction surfaces on outer walls (edge effect)
Material	In steel concrete and steel parts in walls and inadequate thermal insulation of the outer walls, for example, at lintels or balcony plates ("balcony cooling fin")
Mass flow	Placement of noninsulated water pipes or due to building parts
Surroundings	Due to a closet or mattress being flush with the wall or the floor

2.3
Residents' Behavior, Ventilation Behavior

Residents' behavior can also cause situations in which water condensation occurs. Large amounts of water are emitted with normal utilization. The following amounts are released into the air per resident: 1.5–3.0 kg/day by cooking, 2.6 kg/h by showering, 0.7 kg/h by bathing, 1.0–3.5 kg/h by drying laundry, 0.5–1.0 kg/day by watering plants and 1.0–1.5 kg/day by the residents themselves [12]. Since relative humidity is usually lower in outdoor than in indoor air, water can be transported outside by adequate ventilation. Inadequate ventilation can raise the material moisture over the minimal a_w value, inducing corresponding mold growth. The resident can increase the air moisture by

- Inadequate or infrequent airing, no adjustment of airing behavior after renovation.
- Inappropriate airing, for example, permanent airing by tipping the window, thus causing subcooling of adjacent surfaces.
- Airing of cold rooms with warmer outside air.
- Inadequate heating, a too low room temperature (below 16 °C), for example, in bedrooms, heating cold rooms with warm air from heated rooms.
- High humidity (above 60% RH) caused by a high evaporation rate of room plants or inadequate airing and heating.
- Keeping pets in a too small a space.
- Impairment of thermal convection and air circulation by faulty positioning of furniture like beds, wardrobes and drapes.
- Neglecting to clean water drains on balconies.
- Defective maintenance of circulating air units or room air conditioners, no drain in drip pans of circulating air units.
- Reduction in passive air exchange by installing sealed, thermally insulated and/or soundproof windows.

2.4
Genera of Bacteria and Fungi Damaging Building Material

In indoor areas, nearly all building materials can be nutrient sources and thus can be attacked by microorganisms. They include building materials like wallpaper, layers of wall plaster or paint, wall facing, insulation, house dust and "natural" substrates like plant soil, contents of the organic waste container, pet litter or the pets themselves. However, molds are not able to attack mineral-based building materials like bricks, limestone mortar or ceramics, but if the moisture threshold is exceeded, nutrient deposits in the form of fat or dust are usually sufficient for microorganisms to colonize these surfaces. Some xerophilic molds find suitable living conditions even on relatively dry dust. They include certain species of *Aspergillus* (e.g., *Aspergillus amstelodami, Aspergillus halophilus, Aspergillus versicolor*), *Chrysosporium, Wallemia* and *Xeromyces*, which can still grow with a water activity of 0.70 or less. Window frames are frequently colonized by *Alternaria, Aureobasidium* and *Cladosporium*. A very usable substrate is woodchip wallpaper with a high percentage of cellulose, sugar, protein and water-based paints with a sugar-based swelling agent [6]. In damp rooms and especially on damp inner surfaces of outer walls, these materials frequently evidence visible growth of molds from the genera *Penicillium, Cladosporium, Aspergillus, Alternaria, Mucor, Fusarium* and *Aureobasidium*. A large number of molds (*Alternaria, Aspergillus, Chaetomium, Cladosporium, Fusarium, Mucor, Paecilomyces, Penicillium, Trichoderma* among others [6]) grow on urea–formaldehyde foam, which was used for the insulation of older buildings.

Plant soil is also a good substrate for molds whose natural environment is soil. In 29 indoor rooms, *A. fumigatus* was detected in the soil of up to 70% of 181 plants. *Aspergillus* species can also be released from neglected hydroponics [13]. Other sites of molds in interior spaces are listed in Table 4.

Organic waste storage plays a particularly important role in indoor spaces. This is a habitat for thermotolerant molds like *Aspergillus* and *Mucor*. Up to 80% of the fungal flora in organic waste and compost may consist of *A. fumigatus* [6]. Significant microbial emissions may occur in the case of inadequate house hygiene or infrequent emptying of the containers.

Table 4 describes the genera in moldy homes, although it must be taken into consideration that the spectra of the genera may vary in different climate zones.

Thermophilic actinomycetes that start to grow at a temperature of 30 °C may be present indoors on grains, fruit, vegetables and animal litter. Moreover, high concentrations may be found in air-conditioned rooms, when using humidifiers (printer's humidifier lung), in moisture penetration of building materials and in bird and poultry keeping (bird breeder's lung).

Table 4 Mold genera in moldy homes [6]

Indoor area	Mold genera	
Kitchen	Generally	*Penicillium roqueforti* (in refrigerators, on bread cutters, on moldy bread, in trash cans)
Toilet and bathroom	Germany	*Aspergillus flavus, Aspergillus parasiticus*
	Egypt	*Aspergillus, Cladosporium, Penicillium*
	Belgium	*Cladosporium sphaerospermum*
Living room	Belgium	*Aspergillus versicolor, Cladosporium sphaerospermum, Penicillium, Cladosporium herbarum*
	Taiwan	*Penicillium citrinum, Penicillium crustosum, Penicillium implicatum*
Bedroom	Generally	Xerophilic mold genera in household dust: *Eurotium repens, Aspergillus penicilloides, Penicillium chrysogenum, Penicillium brevicompactum, Wallemia sebi, Eurotium halophilum*
	Belgium	*Eurotium repens, Alternaria alternata, Aspergillus penicilloides, Aspergillus versicolor, Aureobasidium pullulans, Penicillium chrysogenum*
Baths and saunas	Generally	*Cephalosporium, Geotrichum, Penicillium, Aspergillus, Chrysogenum, Fusarium, Mucor, Scopulariopsis brevicaulis, Verticillium*
	France	*Alternaria alternata, Cladosporium cladosporioides*
Schools	USA	*Cladosporium, Penicillium, Aspergillus*
Hospitals		*Penicillium*
Skylab		*Aspergillus, Aureobasidium, Cladosporium, Mikroascus, Penicillium, Periconia*

2.5
Visible and Nonvisible Damage

Significant growth of microorganisms, especially molds, may cause adverse health effects. Visible growth is easily detected in most cases. The urgency of remedial actions depends on the extent of damage. According to Ref. [14] infected sites are subdivided into three categories (Table 5). Since symptoms caused by microorganisms are very unspecific ("fungal syndrome" [15]), the presence of hidden mold growth must be excluded, if symptoms of the fungal syndrome occur.

Table 5 The urgency of remedial actions depending on the extent of damage [14]

	Category 1	Category 2	Category 3
Extent of the damage	No or very low bio-mass, for example, slight growth and damage on surfaces smaller than 20 cm^2	Moderate biomass: superficial growth on an area smaller than 0.50 m^2; deeper layers are infected only in a few spots	Large biomass: superficial growth on an area larger than 0.50 m^2; deeper layers may be infected to a high extent
Evaluation	"Normal"	Low-to-moderate damage	Extensive damage
Consequence	As a rule no remedial action is necessary	The emission of microbial products should be suppressed immediately. The causes should be found and eliminated over the medium term	The emission of microbial products should be suppressed immediately. The causes should be found immediately. Remedial action is very urgent. The success of remedial actions must be checked. Medical care by specialists for environ-mental medicine is recommended

3
Measurement Strategy and Differentiation

3.1
Air Sampling, Sedimentation Plates

The detection of airborne spores of fungi and bacteria is quite demanding, since many aspects have to be considered, the most important being complete sampling of all biological airborne particles without discriminating between certain groups of microorganisms. Sampling should be performed in a gentle manner, since most detection methods target cultivable microorganisms. Airborne microorganisms usually weakened by drying or UV radiation may no longer be cultivable after impact sampling, which often operates with air velocities of more than 25 m/s. It is also important that the microorganisms collected do not dry out during the sampling procedure. Most methods sample directly onto nutrient dishes, eliminating further treatment steps which may cause losses or distort the spectrum of the microorganisms sampled. Widely used sampling techniques include impaction and filtration methods, followed by centrifugation, impingement and adsorption in liquid media. The very reliable impaction method can be combined with a multistage sampling technique. There are one-, two-, six- and eight-stage cascade samplers, which subdivide airborne particles into sizes typically ranging from 0.43 to 10 µm (e.g., the Andersen impactor by Andersen Samplers., USA). In a comparison of air samplers, it was demonstrated that Andersen cascade impactors have the highest precision and sampling rates [16]. In this "sieve" impactor, the sample is drawn through approximately 400 holes with the airborne particles subsequently impacting onto petri dishes filled with nutrient agar. Sampling must be carefully timed to avoid overloading at each stage. If the concentration range is unknown, different air volumes must be sampled, which requires different sampling times, since the air flow rate should be constant (typically 1 cubic foot per liter gives 28.3 l/min). The slit sampler applies another type of impactor geometry. These samplers use a turntable which rotates once per sampling interval to distribute the microorganisms over the nutrient agar. Filtering techniques have to ensure that all microbial particles are collected efficiently and that the microorganisms sampled survive quantitatively, if a cultivable method is chosen. Recommended filters consist of water-soluble gelatin, insoluble polycarbonate or cellulose nitrate. The advantage of gelatin filters is that they can be dispersed in dilution fluid or nutrient media, which can then be diluted or transferred to different selective media.

The air sampling technique is commonly used because it enables a toxicological evaluation. Here, the cfu, i.e., spores of fungi or bacteria which have developed into visible colonies, are counted and calculated per cubic meter in relation to the sample volume. The spectrum of the genera detected can also be interpreted in terms of cfu per cubic meter to evaluate the contamination level. Multistage impactors provide information about the penetration depth of biological particles into the trachea and lungs. Air sampling procedures are usually of the short-time type and generally last 2–10 min. Thus, one disadvantage of the air sampling method is the possible occurrence of false-negative results, i.e., the results do not indicate a mold or bacteria problem in the sampling area. This can be explained, on the one hand, by the fact that mold spores are emitted only at certain growth stages. On the other hand, bacteria and mold spores are not equally distributed in indoor air like gaseous volatile organic compounds (VOC) or microbial volatile organic compounds (MVOC) but appear in "clouds" depending on the ventilation procedures and the behavior of the residents or sampling team. Thus, a number of parallel samples (at least three) should be taken, which is very labor-intensive, especially if many different nutrient plates are used for collecting bacteria and different kinds of mold genera. So-called aggressive sampling has been discussed to avoid false-negative results. This means increasing the chance of spore release by knocking on walls and stamping on the ground instead of performing the measurements in a steady state after many hours without ventilation. The air sampling technique may also yield false-positive results when microorganisms from the outdoor air are ventilated into the rooms in question. This is especially true if typical mold damage indicators are airborne owing to nearby emission sources. To minimize this effect, it is recommended that all doors and windows are tightly closed after the last ventilation and that there is a resting period of at least 8 h (preferably 12 h). However, since measurements are usually not automated like particle counting, there is always some air exchange, for example, when the measurement team enters the rooms before and during sampling.

With regard to quality assurance for microbial air sampling, it must be pointed out that laboratory calibration procedures like those for sampling VOCs in indoor air have not yet been adequately developed. It is not possible to determine the actual efficiency of different samplers, since the real concentration of genera distributed in the air is unknown. However, comparative measurements are available for some sampler types.

It should also be mentioned that the majority of microbial airborne particles are dead or "noncultivable", which means that they do not grow under the laboratory conditions chosen. This is, for example, true for microorganisms like *Stachybotrys chartarum*. To fully survey the indoor microbial situation and to avoid overlooking fungal species which may be indicators of damage, additional samples have to be taken using methods like direct microscopy which are not dependent on living, cultivable cultures. Here, for example, microscopic slides with an adhesive surface are inserted in special impactor samplers. The airborne particles collected are stained in the laboratory. Especially spores with a characteristic shape like those of *S. chartarum* are easily detected under the microscope.

Another "quasi" air sampling method is often used to get a snapshot of the indoor air situation. Airborne particles have a certain falling velocity (sedimentation speed) according to their size and spore form. Utilizing this feature, "sedimentation plates", i.e., normal petri dishes with nutrient agar, are laid out throughout the building or indoor area under question. This sedimentation method is very inexpensive because it requires neither sampling equipment nor specially trained personnel. Thus, this method is sometimes preferred by physicians treating patients with unclear "fungal symptoms". However, it should be mentioned that this method does not quantify indoor air concentrations or necessarily qualify the spectrum of the indoor genera, since large spores are overestimated and small ones are often not found at all, if air turbulence prevents sedimentation.

3.2
Household/Carpet Dust

Other sampling techniques are performed alternatively or together with active air sampling methods, since those are sometimes labor-intensive and not always reliable. A common sampling method is the evaluation of household or carpet dust, where the airborne microorganisms are deposited and easily analyzed. Most bacteria and mold genera are able to survive in dust for some time, and microbial growth only occurs with a very high relative humidity (above 97% [17]) or with genera requiring a very low relative humidity (xerophilic species like *Wallemia sebi* or *Aspergillus repens*). Thus, household dust is a fairly good indicator of the microbial indoor air situation and a reliable predictor of microbial damage. Household dust has thus far been sampled with a normal household vacuum cleaner. The dust bag is then shipped to a microbial laboratory. The advantage of using household dust is that a longer period of up to several weeks can be monitored, which increases the chances of being in the sporulating window. Another advantage is that the "dust sampler" is available in nearly every household. For these reasons, this method is very common and usually yields reliable results, although there are some disadvantages. Firstly, it is obvious that a direct toxicological evaluation, i.e., a conversion from cfu per gram to cfu per cubic meter or a calculation of a toxicological risk, cannot be derived from these results. Secondly, the sampling period is normally not defined. Thus, preliminary cleaning with a powerful vacuum cleaner is recommended before taking the sample in order to clearly define the sampling (sedimentation) interval of the microorganisms. There should be a 1-week interval between preliminary cleaning and sampling, since the spectrum of microorganisms in dust can shift within weeks because very sensitive genera die or are no longer cultivable or xerophilic microorganisms may increase in number. Samples should be collected from floors and carpets, since the guidelines for evaluating household dust are based on sampling from these areas. If the sample is taken by untrained persons (i.e., residents), they should be advised that the dust should not be collected from suspicious or obviously contaminated areas like walls. Samples collected by residents are not legally admissible. In the evaluation of household dust, some authors recommend that the sample be sieved to get more stable results [18]. A typical sieving cutoff diameter is either 63 µm or 125 µm. Here, it must be taken into consideration that sieving leads to a higher number of cfu per gram and the appropriate guidelines should be considered when using results from sieved samples. Furthermore, sieving of dust samples may lead to cross-contamination in the laboratory or even contamination of the personnel. Another problem when using vacuum cleaner bags is tube contamination from cleaning other contaminated areas. Thus, collecting samples in household vacuum cleaner bags is no longer recommended.

The so-called ALK sampler offers a solution to the previously mentioned shortcomings. It consists of a simple plastic filter holder, into which a glass-fiber filter (diameter 60 mm) is inserted. The filter holder is attached to the front of a vacuum cleaner tube, which prevents cross-contamination. Since the method is very sensitive, sampling of 1 m^2 is sufficient for normal investigations. The standard protocol consists of marking off 1 m^2 with a tape and thoroughly vacuuming this area with a vacuum cleaner at 1,000 W or more for 5 min. Since a defined sedimentation time is recommended for the evaluation, the marked area should be cleaned 1 week earlier. The residents or room users can carry out the preliminary cleaning. Obviously, the sampling area should no longer be cleaned during the sedimentation period. And it should not be ex-

posed to contamination from other sources, for example, animal cages and pot-
ted plants, but it may be walked upon. After 1 week, the marked area is sampled
with the ALK sampler for 5 min. In the laboratory, the filter is suspended in
physiological NaCl solution and shaken under defined conditions. The sus-
pension is diluted according to the expected contamination level and plated out
on petri dishes. Cfu distributions and distributions of genera in carpet or
household dust samples concerning microbially contaminated and nonconta-
minated indoor areas will soon be published by the authors. With the results of
the household dust sampling the indoor areas in question can be classified into
moldy or nonmoldy homes with a specified error.

3.3
Surfaces and Building Material

Contact samples or samples of the building material itself should be obtained
to determine whether building materials are infected or to toxicologically evaluate the genera causing the microbial damage. Contact samples are method-
ologically subdivided into cultivable/viable contact culture samples and nonviable samples. Petri dishes filled with nutrient agar over the rim of the dish are
usually used for the viable culture samples. This type of petri dish is known as
a recovering organisms detecting and counting dish (RODAC). The RODAC
dish is briefly pressed onto the area in question and then closed and tightly
sealed immediately to avoid contamination during transport. Bacteria and
mold spores stick to the nutrient agar and start to grow under the appropriate
incubation conditions. Since the surface area of the RODAC dish is small (usually 20 cm^2), several samples have to be taken in order to get a representative
overview of the damage. The disadvantage of this technique consists in the high
probability that the nutrient agar is overgrown by fast-growing molds, so other
molds which could be used as damage indicators cannot develop into visible
colonies. This is especially true for samples from microbial damage. A qualitative result often cannot be obtained with this method, even if the nutrient
dishes are periodically checked for growth and the spectrum of genera. The additional application of the nonviable technique is recommended to circumvent
this disadvantage and to get quick results. Here, a clear adhesive tape is pressed
onto the building material in question, removed and transferred to a microscope slide. The tape sample is then conventionally stained and can be micro-
scopically analyzed right away. If a microscope is available at the contaminated
site, samples can be evaluated there. It should be noted that vertical, smooth
surfaces like walls are never sterile but always show airborne microorganism
contamination. Thus, it must be carefully determined whether the sample represents real microbial growth or a normal background level of mold spores or
bacteria.

3.4
Viable/Nonviable Counts, Nutrients, Differentiation of Fungi and Bacteria

Despite the rapid development of molecular biological, immunological and chemotaxonomic techniques, molds from field samples are still differentiated largely by classical microscopic evaluation [4, 19]. Morphological characteristics of hyphae and mycelia can be visualized by different stainings. This can be a very time consuming process depending on the knowledge of the investigator or the use of genera and species keys. Moreover, samples are examined using different kinds of culture media, in which molds find conditions matching their physiological demands [20]. Thus, the main advantage of culturing on nutrient agars is that the genera and species of molds can normally be differentiated. Unfortunately molds cannot always be induced to produce spores under laboratory conditions, though this is essential for differentiation. When molds do not sporulate and for this reason cannot be differentiated, "sterile mycelium" is indicated. The lower the percentage of sterile mycelium, the greater the expertise of the laboratory. However, it must be said that inducing spore formation may take weeks. The main disadvantages of differentiating viable microorganisms are that the procedure is time-consuming and requires laboratory capacity and skilled personnel. This is especially true for molds, since they are mainly differentiated after microscopic evaluation with the use of illustrated handbooks. A variety of nutrient agars have to be used depending on the different requirements of the bacteria and molds, since a universal nutrient agar does not exist. Parallel sets of nutrient agars have to be inoculated owing to the different incubation temperatures required (22±1 °C for mesophilic, 36–37 °C for thermotolerant and 56 °C for thermophilic microorganisms).

Sometimes spores indicating mold damage, the so-called indicator molds, are not cultivable. Thus, nonviable techniques prove to be particularly advantageous, since typical spores like those of S. chartarum are easily recognizable under the microscope. Small spores, nonexpressive ones and bacteria cannot be differentiated by noncultivable means. Experience has shown that noncultivable spores and bacteria greatly exceed "living", cultivable microorganisms, sometimes by 100–10,000 times. This has to be taken into consideration when evaluating the successful rehabilitation of an indoor environment or giving an expert opinion on the allergenic potential, since spores reveal their allergenic potential in all states ("dead and alive"). This phenomenon is especially true for S. chartarum. Here, air samples on culture media showed negative results even on cellulose-based nutrient agar, while microscopy yielded 7,500 spores of S. chartarum per cubic meter [20]. Additionally special staining techniques allow differentiation between the sum of all bacteria or mould spores from active ones [21].

Bacteria can easily be detected by sampling on nutrient agar containing blood. Bacillus spores are often found in normal air. In contrast to molds, whose spores serve for proliferation via the air and are very susceptible like all other viable forms, only the genera *Bacillus* and *Clostridium* produce spores in hostile surroundings, for example, on nutrients. Bacterial spores are very resistant to drying out and especially to different thermal conditions. This is why they can be found everywhere without any significant effect on health. Another typical airborne bacterial species is *Micrococcus luteus*, which forms yellow colonies on blood agar. Many airborne bacteria form colored colonies, especially on blood agar. Their coloring provides greater protection against permanent outdoor UV radiation for prolonged survival in the air. However, the eye-catching colors on blood agar do not indicate the facultative or obligate pathogenicity of these species. Even hemolytic properties, i.e., the destruction of the red color of blood agar, does not necessarily correlate with any pathogenic properties or health effects.

So-called Gram staining is a more reliable determination method. Bacteria staining red are deemed "Gram-positive" and are often typical airborne microorganisms. This does not mean that they grow in the airborne state but that they are more protected against hostile conditions than Gram-negative bacteria. Gram-negative bacteria, which appear blue under the microscope, are normally not found in the air or only at a very low level. Their detection in air is a strong indication that there is "real" water in the building. Typical water bacteria are of the genus *Pseudomonas*, a pathogenic species being *Pseudomonas aeruginosa*. Molds usually start to grow with a high relative air humidity or relatively high water content, while more bacteria are found on wet materials than molds.

3.5
Genetic Fingerprinting for Detecting Bacteria and Molds

Genetic fingerprinting has emerged as a fast and reproducible technique for detecting bacteria and some molds down to the species level such as *A. fumigatus*. This new technique analyzes genetic markers to identify and check taxonomic relationships. A number of different methods are used, including multilocus enzyme electrophoresis, restriction fragment length polymorphism, randomly amplified polymorphic DNA analysis, electrophoretic karyotyping and sequencing [22–24]. Genetic methods are common and accepted for identifying bacteria [25]. There exists a variety of polymerase chain reaction (PCR) methods for analyzing molds, but the most common ones examine the 18S rDNA and 28S rDNA sequences as well as the internal transcribed spacer regions ITS I and ITS II [26–29]. In mycological laboratories, research has been conducted on pathogenic fungi and yeasts like *A. fumigatus* and *Candida albicans* [30–34]. In recent years, approaches have also been developed for detecting molds in environmental samples and food, like milk, flour and grains [35–37]. However, there are very few validated standards for the detection of molds with these fingerprinting methods in clinical or environmental labs. Finally, PCR methods must be significantly adapted and validated before they can be used in routine analysis [38, 39]. There are only a few studies on quality assurance which compare PCR, immunological techniques like Western Blot or enzyme-linked immunosorbent assay or analytical methods like gas chromatography for mold detection [40, 41].

3.6
Enzymatic Tests for Mold Growth

To detect and quantify molds on surfaces, the University of Copenhagen developed a method based on an enzymatic reaction [42]. In this study, a direct correlation was found between the amount of fungal enzymes and the fungal biomass. The enzyme amount is quantified by a fluorescence photometer. The measured units are classified as follows: less than 25 for dust-free surfaces, less than 450 for dusty surfaces without molds, 180–600 for areas with non-viable dried-out mold mycelium and 800–14,000 for surfaces with active mold growth. The sensitivity is reported to be very high (1×10^{-9} g biomass). For example, one spore of *Geotrichum candidum* has a mass of 50×10^{-9} g [43]. This method is regarded as useful in monitoring rehabilitation procedures after mold damage.

3.7
Search for Hidden Mold Growth

3.7.1
Microbial Volatile Organic Compounds as Indicators for Hidden Mold Growth, Selection of Microbial Volatile Organic Compounds, Analysis, Evaluation

Hidden microbial growth, especially mold growth, cannot always be reliably detected with microbial methods. For example, microbiological air sampling may lead to false-negative results owing to nonsporulating molds or owing to the fact that the molds are prevented from getting airborne. One alternative is a chemical-analytical method, since molds and bacteria produce a variety of VOC, the so-called MVOC, during all growth periods. These MVOC evaporate from the biomass into the indoor air, where they are equally distributed by diffusion processes like all other gaseous VOC. In the meantime, MVOC measurements have been widely used as indicators for hidden mold growth [14, 44, 45].

The analytical technique is comparable to VOC measurements. Using a vacuum pump, an air sample is drawn through a sorbent, where the MVOCs are quantitatively absorbed. The analysis is achieved by gas chromatography/mass spectrometry. There are two ways of collecting the sample and transferring it to the analytical system. One method uses activated charcoal as a sorbent and a solvent or solvent mixture for eluting the MVOC [46–48]. Here, a relatively large air sample is taken to achieve a detection limit in the range 0.1–0.5 µg/m³. The second method uses Tenax TA as a sorbent, and transfer of the sample to the gas chromatography/mass spectrometry system

Table 6 Overview of the sampling and the desorption techniques

Sampling	Sampling and desorption technique	
	On activated charcoal	On Tenax TA
Desorption	By a solvent (mixture)	By thermal energy
Laboratory demands	Moderate to high	High
Blank values/contamination	Low risk	High risk
Aliquot of sample injected	1:4000–1:1000	1:10–1:1 (=complete)
Multiple injection	Possible	Impossible
Sensitivity	High	Very high
Detection limit	0.1–1.0 µg/m^3	0.01–0.1 µg/m^3
Sample volume (l)	100–500	2–5
Ecological and health aspect of solvents	Risky owing to the use	No solvent needed

is achieved by thermal energy alone. Each method has its specific advantages (Table 6).

Lower (i.e., better) detection limits are achieved with the thermal desorption method owing to the higher aliquot transferred to the analytical device. The shortcoming of the charcoal method is the injection of only a very small part (aliquot) of a sample, which can only be partly compensated by sampling a larger air volume. The disadvantage of the thermal desorption technique is that quality assurance requires the collection of at least duplicate samples. Another disadvantage of this method is the need for special equipment and specially trained personnel.

It has been suggested that certain MVOC are useful indicators of hidden microbial damage [49–56]. A scheme was developed for interpreting the analytical results of indoor MVOC concentrations [45, 57].

According to Ref. [45] primary indicators are 3-methylfurane, 1-octen-3-ol, dimethyl disulfide and sometimes 3-methyl-1-butanol, and secondary indicators are 2-hexanone and 2-heptanon (Table 7).

According to Ref. [57] primary indicators include 2-methylfurane, 3-methylfurane, dimethyl sulfide, dimethyl disulfide, dimethyl sulfoxide, 2-pentanol, 1-octen-3-ol, 2-methylisoborneol and geosmine, and secondary indicators are 2-3-methyl-1-butanol, 3-methyl-1-butanol, 2-methyl-1-propanol, 1-decanol, 3-octanol, 3-octanone and 2-heptanone. The higher the number of indicators detected, the more reliable the interpretation (Table 8).

It must be pointed out, however, that the use of MVOC as indicators for hidden mold damage has not been scientifically evaluated. The most crucial point is that nearly all MVOC are not only produced by microbes but also have

Table 7 Interpretation of microbial volatile organic compound (*MVOC*) results according to Lorenz [45], modified by Landesgesundheitsamt Baden-Württemberg [58]. *Sum of MVOC* is the arithmetical sum of primary and secondary indicators

µg/m³	No detection of a primary indicator	Primary indicator in the range between 0.05 and 0.10 µg/m³	Detection of at least one primary indicator above 0.10 µg/m³
Sum of MVOC ≤ 0.50	No hidden microbial damage	Limited microbial damage possible	Microbial damage possible
0.50 > sum of MVOC ≤ 1.0	Probably no hidden microbial damage	Microbial damage possible	Microbial damage very probable
Sum of MVOC >1.1	Microbial damage possible	Microbial damage very probable	Microbial damage is proven

Table 8 Interpretation of MVOC results [57]

Interpretation	Sum of MVOC of primary indicators (µg/m³)	Sum of MVOC of primary and secondary indicators (µg/m³)
Low indication	0.05–0.10	<0.20
Moderate indication	0.01–0.30	0.20–0.50
Clear indication	0.30–0.50	0.50–1.0
Obvious indication	0.50–1.0	1.0–1.5
Substantial indication	>1.0	>1.5

other sources. Products used in buildings such as solvents, paints, and adhesives as well as new furniture and furnishings release analogous VOC. Other sources are well known, like tobacco smoke and food. Some nonbiogenic sources for compounds which are also produced by microorganisms are given in Table 9 [59–65].

Another problem is that other indoor VOC coelute chromatographically at the same retention time, thus disrupting the findings of mass spectrography. Since indoor VOC regularly occur in concentrations 10–1,000 times higher than MVOC, the misinterpretation of mass spectrographs is likely when only using the single ion monitoring mode.[1] Therefore, an analysis of the VOC spectrum in the full scan mode is strongly recommended. The third point is that only a relatively high source strength of MVOCs (in micrograms per hour) may yield measurable results in indoor air. Surprisingly low MVOC emission

---

[1] In order to increase sensitivity, only some but not all tracks of fragments are monitored.

Table 9 Volatile organic compounds (*VOC*) (nonmicrobial) sources of selected indicator MVOC. (Artificial) flavoring (*1*), beer (*2*), cauliflower (boiled) (*3*), coating materials (*4*), fat (*5*), cream components (*6*), essential oil (*7*), paints (*8*), coffee (*9*), coffee flavoring (*10*), cabbage (*11*), coconut fat (*12*), varnishes (*13*), solvents (*14*), perfumes (*15*), leeks (cooked) (*16*), cleaning agents (*17*), air freshener (*18*), ointment (*19*), shellfish (cooked) (*20*), chives (*21*), tobacco smoke (*22*)

Indicator MVOC	Nonbiogenic indoor VOC sources
2-Methyl-1-propanol	1, 8, 13, 14, 17, 18
2-Methyl-1-butanol	13, 14
3-Methyl-1-butanol	13, 14
2-Pentanol	8, 13, 14
1-Octene-3-ol	1 (sea fish, crabs), 7 (lavender, mint)
7-Octene-2-ol	Unknown
3-Octanol	Unknown
1-Decanol	6, 15, 19
2-Heptanone	5, 12, 13, 14
3-Octanone	13, 15
2-Methylisoborneole	1, 10
2-Methylfurane	22
3-Methylfurane	22
Dimethyl sulfide	3, 16, 20, 21
Dimethyl disulfide	1 (beer, coffee, cabbage)
Dimethyl sulfoxide	4 (rare)

rates by molds growing on building material were reported by Schleibinger at al. [66]. If the infected area and air exchange rate per hour is given, expected concentrations in indoor air can be taken from Table 10. Cautious calculations based on laboratory trials suggest that the concentrations provoked by microorganisms may be below the analytical detection limit in indoor air. Research is being done in moldy and nonmoldy homes to clarify this point. Moreover, laboratory experiments must provide more knowledge about how MVOC production is related to the substrate and genera [66]. Obviously, MVOC measurements should not be the only method used to detect hidden microbial damage; however, hidden microbial damage in cavities, especially from molds, might be detected by the MVOC method, since the air exchange rates are very low in these spaces. Here, the thermal desorption method is most advantageous because it usually requires only 2–3 l of sample volume.

Table 10 Calculated indoor air concentrations from laboratory experiments. Infected site 0.25 m². Air exchange rate (AER) 0.1, 0.2, 0.5 and 1.0 1/h. Room volume 50 m³. Time elapsed after ventilation and closure of the room 8 h

Emission rate in laboratory incubation chambers	Calculated emission rate based on a defined area	Calculated indoor air concentrations by a mold-infested site of 0.25 m²			
Area 20 cm² (µg/week)	Area 0.25 m² (µg/h)	AER=0.1 (µg/m³)	AER=0.2 (µg/m³)	AER=0.5 (µg/m³)	AER=1.0 (µg/m³)
0.001	0.00074	0.000082	0.000059	0.000029	0.000015
0.005	0.0037	0.00041	0.00030	0.00015	0.000074
0.01	0.0074	0.00082	0.00059	0.00029	0.00015
0.05	0.037	0.0041	0.0030	0.0015	0.00074
0.1	0.074	0.0082	0.0059	0.0029	0.0015
0.5	0.370	0.041	0.030	0.015	0.0074

3.7.2
Odor Detection by the Mold Sniffer Dog and Humans

Since 1999 specially trained dogs, so-called mold sniffer dogs, have been used for the rapid detection of hidden mold damage. They detect damage in areas not only at their level but also on indoor ceilings. Their usefulness has not been scientifically evaluated, but there are reports on their successful use in such procedures [67]. No decisions about rehabilitation should be made solely on the basis of detection by the sniffer dog [14]. Odor perceived by humans is not a quantitative measure for microbial damage because the quantity of typical odorous substances does not necessarily correlate with the biomass or the area

Table 11 Odor thresholds of typical MVOC

Compound	Odor threshold (µg/m³)	Odor quality
2-Methyl-1-propanol	3	Musty, moldy
2-Methyl-1-butanol	45	Sour, biting
3-Methyl-1-butanol	30	Foul smelling
1-Octen-3-ol	16	Moldy
2-Heptanone	94	Fruitlike
2-Methylisoborneol	0.007	Earthlike
Dimethyl sulfide	2	Disagreeable
Dimethyl disulfide	0.1	Sulfurous, cabbage-like
Geosmin (trans-1,10-dimethyl-trans-9-decalol)	0.1	Earthlike

infested by the microorganisms [68]. The constellation of only a few compounds with a low odor threshold is decisive for odor perception. Olfactory measurements showed that the earthlike smell of geosmine or the mold smell of 2-methylisoborneol is masked by other odorous compounds, leading to a completely different odor perception [69]. The odor thresholds of typical MVOCs are given in Table 11 [43, 51, 53, 59–65, 70–73].

4
Rehabilitation of Microbial Damage in an Indoor Environment

Basically microbial growth can only be prevented by a consequent elimination of the dampness and/or water entering the building, since the RH and the moisture content are normally the limiting factors (see Sect. 2). Rehabilitation procedures have to be preceded by a careful investigation and a thorough understanding of the causes of the microbial growth in the indoor area. If the causes of dampness are not addressed, the microbial growth will most likely reappear within weeks or months. Rehabilitation with chemicals is only recommended under certain circumstances, since many agents are only bacteriostatic or fungistatic. Thus, the microbial growth will redevelop within a short time after the chemical concentration has decayed below the bacteriostatic or fungistatic concentration. In many cases, an additional burden by volatile compounds is caused by antimicrobial chemical treatment, especially if the damage was due to a poor (and still unimproved) ventilation rate. Although the chemicals in question are usually not antigens, unspecific or allergic-type reactions are sometimes observed after applying the chemicals. Thus, the observed fungal-syndrome-like symptoms often seem to increase or, at least, do not disappear.

Concerning the rehabilitation principle one must always bear in mind that "only a removed microorganism is a good microorganism", which places the emphasis on removing the biomass and not just killing the microorganisms in question. This is because the main problem is not the infection per se but the fact that, for example, the mold spores keep their allergenic potential even if killed (see Sect. 5). Moreover, destroying microorganisms, especially fungal spores, may release mycotoxins from the spore interior. Microorganisms are removed from hard surfaces by normal washing procedures, possibly with a detergent. Disinfectants are not necessary in this case, since the aim is the quantitative removal from the building. Microorganisms are no longer a risk as soon as they have been removed from the indoor space in a plastic bag. If microbial growth, especially of the mold mycelium, has grown deeper into wallpaper or rough building materials, those materials have to be completely and carefully removed.

Microbial growth on relatively small surfaces (below 0.50 m^2) can be removed by taking normal precautions (gloves, mask with an efficient filter), and if the persons have no special risks, like allergies. Larger surfaces and building material, which is infected to a greater depth, should be removed by trained experts who are familiar with the local regulations, industrial safety procedures and the removal of biological material. On surfaces like plaster, a layer of several millimeters should be removed even if only superficial growth is detected. All plaster material should be removed in the case of deep cracks with water coming from outside areas. All procedures like polishing or cleaning surfaces should by accompanied by high efficiency vacuuming and high-efficiency particulate air filtering. Other parts of the buildings have to be hermetically sealed. The proliferation of bacteria and mold spores has to be rigorously avoided for health reasons, since enormous spore concentrations are released into the air during rehabilitation. The rooms should have a negative air pressure whenever possible to prevent air flowing into uncontaminated areas. After removal and refurbishment, the materials should be given sufficient time to dry out. The moisture content of the building material should be measured to guarantee the success of the rehabilitation procedures.

5
Hygienic and Toxicological Evaluation of Microorganisms in an Indoor Environment

At present there is no binding or mandatory regulation concerning the evaluation of the presence or the growth of bacteria and moulds indoors and the danger thereby encountered owing to those microorganisms, as clear dose/response relationships seem not to exist. However, bacterial and mold growth must be suppressed in indoor areas for hygienic and toxicological reasons, since microbial particles may cause different individual reactions and should be properly removed whenever detected. Many futile attempts have been made to calculate the risk posed by microorganisms. Even if it were possible to precisely measure the biomass of existent microbial damage, it would be very difficult to determine the exact risk and toxicological significance. A rough estimation of the microbial damage, expressed as an area, is often used as a crude substitute for the "dose" of dangerous microbial particles. In this context, the urgency of the rehabilitation procedure depends on the size of the infected site (Table 5). Apart from questions as to the urgency and thoroughness of removing visible microbial damage, methods for locating hidden microbial damage are also under discussion. This is of interest as there are cases of complaints by users of flats and offices in which microbial growth was found only after months and years. In this context the causes for symptoms of the fungal syndrome are ignored too often.

5.1
Infections by Bacteria and Molds

Fortunately it is very rare that infections are caused by inhaled airborne mold spores or bacteria resulting from microbial damage. Only a few mold species are considered facultative pathogens for humans and even fewer obligatory pathogens. Many of the pathogenic molds are ubiquitous saprophytes that may, for example, populate the skin and mucous membranes of humans without affecting them in any way. This is generally the case for people with an efficient immune defense. However, molds may be pathogenic in people with a high immunodeficiency who must behave very carefully and follow all hygienic precautions and medical advice. These individuals should be protected by a ventilation system with high-efficiency particulate air (HEPA) filtration and usually have to be hospitalized. The hospital wards should have a positive air pressure, so that the air flows from the patients' rooms into the neighboring rooms, thus protecting the patients from contaminated air. Particularly feared are fungal infections (mycosis), which can be caused by such molds like *A. fumigatus*. The lung is the main site of infection, but the microorganisms may spread from there to other internal organs. Infections are also caused by the following species: *Aspergillus flavus*, *Aspergillus niger*, *Aspergillus terreus*, *Mucor*, *Rhizomucor*, *Rhizopus oryzae*, *Alternaria*, *Fusarium* and *Cunninghamella*. See Ref. [74] for a description of infections caused by molds.

5.2
Toxic (Nonallergenic) Effects

The inhalation of spores, fragments of hyphae or airborne microbial metabolites attached to airborne particles, especially at elevated levels, may lead to unspecific complaints and symptoms. This may also be the case for nonallergic persons. Reactions like (allergic) bronchitis and chronic obstructive pulmonary disease as well as the aggravation of asthma have been observed. Studies by Johanning [15] showed a correlation between damp and/or moldy homes and symptoms like headache, fatigue, burning eyes, blocked or runny nose, lack of concentration, hoarseness, malaise, coughing, wheezing and susceptibility to infections. However, clear dose relationships have been not found here either. Information on toxic effects due to biological particles has been compiled for specific branches like agriculture, intensive mass animal farming, the waste industry and wood processing. But in these industry branches the concentrations of microbial particles are higher by some orders of magnitude and the extrapolation to low-level indoor air situations is very difficult, mostly impossible.

Fungal spores also contain metabolites like mycotoxins [58]. These mycotoxins possess many kinds of toxic and/or cancerogenic properties. They are immunosuppressive, hemorrhagic, hepatotoxic, nephrotoxic and may cause tremors. Many molds growing on nutrient foods like cheese and bread produce very harmful mycotoxins which are released into the foodstuff. These effects have been well documented, for example, the production of aflatoxins on peanuts by the mold species *A. flavus*. But it has not yet been examined whether mycotoxin production plays an important role if molds grow on building materials and the spores are then inhaled. Since there is presumably no lowest effect level for carcinogenic compounds, these effects should not be ignored.

1,3-β-Glucan, a polymer of D-glucopyranose, is a constituent of the cell wall of molds but also occurs in some genera of bacteria and even in plants and pollen. This compound has toxic properties, like immunosuppressive and inflammatory effects. Since a correlation was found between 1,3-β-glucan and adverse health effects in some studies, this compound might be used as a marker for sick building syndrome, although a causal connection has not been demonstrated.

After the discovery of MVOC, these compounds were also thought to provoke certain symptoms of sick building syndrome in view of their similarity to commonly found VOC, which may cause symptoms like headache and dizziness. However, since MVOC concentrations are normally 2–3 decimal powers lower than those of comparable VOC, these low concentrations should not be expected to have any effects in indoor air [75]. Mutagenic, cytotoxic or genotoxic effects were not found in vitro in mammalian cells with typical MVOC like 2-pentylfuran, 1-butanol, 2-butanone, 3-methyl-1-butanol, 2-methyl-1-propanol, 2-hexanone, 2-heptanone, 1-octen-3-ol, 3-octanone or 1-decanol [76]. In a literature survey no adverse health effects caused by MVOC in concentrations typically encountered in moldy homes were reported [77].

5.3
Allergenic Effects

Many authors attributed allergenic effects to mold exposure in indoor air [53, 78–80]. From the medical point of view, the onset of an allergy almost always depends on the predisposition of the person rather than the frequency or the level of fungal spores. Therefore a clear dose/response relationship cannot be identified. Thus, there are no evidence-based guidelines for the fungal spore concentrations in indoor air or for the spectrum of the microorganisms. Here, it must be strongly emphasized that not only "dead" cells, i.e., microorganisms which are disinfected or no longer cultivable for other reasons, but also microorganism fragments may keep their allergenic properties. Sampling of both viable and nonviable microbial particles must be performed to assess the allergenic potential. Unfortunately, biomonitoring methods for assessing microbial exposure have not yet been fully evaluated. Some studies have reported a correlation between mold exposure and specific antibodies, but too many false-negative or false-positive results have been found thus far.

5.4
Asthma

Asthma or more accurately "occupational or environmental-type asthma" [81] is of great interest for public health, although the specific pathologic mechanisms are still under question. Causality is very difficult to prove here as well [15]. In the USA, 12 million people were found to have asthma in 1998, with a steadily increasing trend. Occupational-type asthma is the disease most commonly associated with the workplace. More than 450 compounds occur at the workplace and are known to trigger asthma. Among these compounds, many are of biological origin, including a broad spectrum of mold species. Although environmental asthma, especially the indoor-related type, is also very important, the causality is even less understood and more difficult to prove.

5.5
Odor-Induced Irritation

Some MVOC have a very bad smell as well as a very low odor threshold (Table 11). Some of these compounds smell musty, moldy or foul. Even if the toxicological effects of MVOC have to be negated at the present time, unpleasant smells are psychologically associated with unhealthy living conditions. Moreover, the smells can provoke nausea and vomiting. Therefore, smelly homes are not acceptable, whether the smell is caused by microorganisms or by other sources.

Acknowledgements We would very much like to thank Anke Siebert for her contribution to Sects. 1.1 and 3.5.

References

1. Lacey J (1994) In: Singh J (ed) Building mycology. Management of decay and health in buildings. E & FN Spon. Chapman and Hall, London, p 77
2. Solomon WR (1975) J Clin Allergy Clin Immunol 56:235
3. Ebner MR, Haselwandter K, Frank A (1992) Mycol Res 96:117
4. Piecková E, Jesenská Z (1999) Ann Agric Environ Med 6:1
5. Tabak HH, Cooke WMB (1968) Bot Rev 34:126
6. Reiss J (1986, 1997) Schimmelpilze – Lebensweise, Nutzen, Schaden, Bekämpfung. Springer, Berlin Heidelberg New York
7. Yang CS (1996) Fungal colonization of HVAC fiber-glass air-duct liner in USA. In: Proceedings of the 7th international conference on indoor air quality and climate, Toronto, p 3:173
8. Scott WJ (1957) Adv Food Res 7:83
9. Mossel and Ingram (1955) The physiology of the microbial spoilage of foods. J Appl Bacteriol 18:232
10. ISIAQ (1996) Control of moisture problems effecting biological indoor air quality. ISAQ international society of indoor air quality and climate. Ottawa, p 8
11. Corry J (1987) Food Beverage Mycol 51
12. Schrodt J (1997) In: Keller R (ed) Gesundheitliche Gefahren durch biogene Luftschadstoffe. Mikrobiologie und Hygiene der Medizinischen Universität zu Lübeck, p 1:11
13. Staib F (1982) Zbl Bakt I Abt Orig B 176:142
14. Umweltbundesamt (2002) Leitfaden zur Vorbeugung, Bewertung und Sanierung von Schimmelpilzwachstum in Innenräumen, Berlin
15. Johanning E (2001) Bioaerosols, fungi and mycotoxins: health effects, assessment, prevention and control. Update and revised edition by Dr med Eckardt Johanning. Fungal Research Group, Albany, NY
16. Verhoeff AP, Wijien JK, Fischer P, Brunekreef B, Boleig JSM, van Reenen ES, Samson RH (1990) Toxicol Ind Health 6:133

176 H. Schleibinger et al.

17. Pasanen AL, Lappalainen S, Korpi A, Pasanan P, Kalliokoski P (1996) Volatile metabolic products of moulds as indicators of mould problems in buildings. In: Proceedings of the 7th international conference on indoor air and climate. Nagoya, p 2:669
18. Baudisch C, Sadek H, v. Stenglin M (2001) Umweltmed Forsch Prax 6:265
19. Petrini LE, Petrini O (2002) Schimmelpilze und deren Bestimmung. Cramer, Berlin
20. Dill I, Trautmann C, Szewzyk R (1997) Mycoses 40:110
21. Kepner RL Jr, Pratt JR (1994) Microbiol Rev 58:603
22. Soll DR (2000) Clin Microbiol Rev 13:332
23. Edel V (1998) Chemical fungal taxonomy. Dekker, New York
24. Williamson EC, Leeming JP (1999) Mycoses 42 Suppl 2:7
25. Vandamme P, Bot B, Gillis M, de Vos P, Kersters K, Swings J (1996) Microbiol Rev 60:407
26. Voigt K, Cigelnik E, O'Donnell K (1999) J Clin Microbiol 37:3957
27. Schmidt O, Moreth U (1998) Holzforschung 52:229
28. Moreth U, Schmidt O (2000) Holzforschung 54:1
29. Chen RS, Tsay JG, Huang YF, Chiou RY (2002) J Food Prot 65:840
30. Schonian G, Tietz HJ, Thanos M, Graser Y (1996) Mycoses 39 Suppl 1:73
31. Einsele H, Hebart H, Roller G, Loffler J, Rothenhofer I, Muller CA, Bowden RA, van Burik J, Engelhard D, Kanz L, Schumacher U (1997) J Clin Microbiol 35:1353
32. Ferrer C, Colom F, Frases S, Mulet E, Abad JL, Alio JL (2001) J Clin Microbiol 39:2873
33. Luo G, Mitchell TG (2002) J Clin Microbiol 40:2860
34. Pham AS, Tarrand JJ, May GS, Lee MS, Kontoyiannis DP, Han XY (2003) Am J Clin Pathol 119:38
35. Vaitilingom M, Gendre F, Brignon P (1998) Appl Environ Microbiol 64:1157
36. Shapira R, Paster N, Eyal O, Menasherov M, Mett A, Salomon R (1996) Appl Environ Microbiol 62:3270
37. Bluhm BH, Flaherty JE, Cousin MA, Woloshuk CP (2002) J Food Prot 65:1955
38. Vaneechoutte M, Van Eldere J (1997) J Med Microbiol 46:188
39. Klont RR, Meis JF, Verweij PE (2001) Clin Microbiol Infect 7 Suppl 2:32
40. Guillot J, Sarfati J, de Barros M, Cadore JL, Jensen HE, Chermette R (1999) Vet Rec 145:348
41. Reiss E, Obayashi T, Orle K, Yoshida M, Zancope-Oliveira RM (2000) Med Mycol 38 Suppl 1:147
42. Reeslev M, Miller M (2000) The MycoMeter-test. A new rapid method for detection and quantification of mold in buildings. In: Proceedings of healthy buildings, p 1:589
43. Miller M, Palojorvi A, Rangger A, Reeslev M, Kjoller A (1998) Appl Environ Microbiol 64:613
44. Keller R, Sönnichsen R (1997) Umweltmed Forsch Prax 2:265
45. Lorenz W (2001) Handbuch für Bioklima und Lufthygiene. Ecomed, Landsberg
46. Sagunski H (1997) Umweltmed Forsch Prax 2:95
47. Schleibinger H, Wurm D, Möritz M, Böck R, Rüden H (1997) Zbl Hyg 200:137
48. Fedoruk MJ, Uhlmann S, Baker DB, Yang H (1999) Bioaerosols, fungi and mycotoxins: effects, assessment, prevention and control. Eastern New York Occupational and Environmental Health Center, Albany, NY
49. Rivers JC, Pheil JD, Wieer RW (1992) J Exposure Anal Environ Epidemiol 1:177
50. Börjesson T, Stöllman U, Schnürer J (1993) J Agric Food Chem 41:2104
51. Wessen B, Ström G, Schoeps KO (1994) MVOC-profiles – a tool for indoor-air quality assessment. Workshop: Indoor air – an integrated approach. Gold Coast, Australia, p 67
52. Batterman SA (1995) In: Burge HA (ed) Bioaerosols. Lewis, Boca Raton, FL, p 249
53. Dewey S, Sagunski H, Palmgren U, Wildeboer B (1995) Zbl Hyg 197:504
54. Sunesson AL, Vaes WHL, Nilsson C-A, Blomquist G, Andersson B, Carlson R (1995) Appl Environ Microbiol 61:2911

55. Pasanen AL, Lappalainen S, Korpi A, Pasanan P, Kalliokoski P (1996) Volatile metabolic products of moulds as indicators of mould problems in buildings. In: Proceedings of the 7th international conference on indoor air and climate. Nagoya, p 2:669

56. Morey P, Worthan A, Weber A, Horner E, Black M, Müller W (1997) Microbial VOCs in moistures damaged buildings. In: Proceedings of healthy buildings–IAQ '97. Washington, DC, p 1:245

57. Keller R (2002) Microbial volatile organic compounds (MVOCs) in Innenräumen: Entwicklung einer Methode zur Detektion von MVOCs aus Schimmelpilzen. Fortschritt-Berichte VDI Reihe 17 Biotechnik/Medizintechnik (19). VDI, Düsseldorf

58. Landesgesundheitsamt Baden-Württemberg (2001) Schimmelpilze in Innenräumen-, Nachweis, Bewertung, Qualitätsmanagement

59. Dellweg H, Schmid RD, Trommer WE (1995) Römpp-Lexikon Biotechnologie. Thieme, Stuttgart

60. Eisenbrand G., Schreier P (1995) Römpp-Lexikon Lebensmittelchemie. Thieme, Stuttgart

61. Falbe J, Regitz M (eds) (1996–1999) Römpp-Lexikon Chemie. Thieme, Stuttgart

62. Steglich W, Fugmann B, Lang-Fugmann S (1997) Römpp-Lexikon Naturstoffe. Thieme, Stuttgart

63. Zorll U (1998) Römpp-Lexikon Lacke und Farben. Thieme, Stuttgart

64. Pühler A, Regitz M, Schmid RD (1999) Römpp-Lexikon Biochemie und Molekularbiologie. Thieme, Stuttgart

65. Hulpke H, Koch HA, Niessner R (eds) (2000) Römpp-Lexikon Umwelt. Thieme, Stuttgart

66. Schleibinger H, Brattig C, Mangler M, Samwer H, Laussmann D, Eis D, Braun P, Marchl D, Nickelmann A, Rüden H (2002) Microbial volatile organic compounds (MVOC) as indicators for fungal damage. In: Proceedings of the 9th international conference on indoor air quality and climate. Monterey, p 4:707

67. Böge K-P (2002) In: Moriske H-J, Turowski E (eds) Handbuch für Bioklima und Lufthygiene. Ecomed, Landsberg

68. Böck R, Schleibinger H, Rüden H (1998) Umweltmed Forsch Prax 3:359

69. Larsen TO, Frisvad JC (1995) Mycol Res 99:1153

70. Miller JD, Laflamme AM, Sobol Y, Lafontaine P, Greenhalgh R (1998) Int Biodeterior 24:103

71. Reynolds S-J, Streifel AJ, McJilton CE (1990) Am Ind Hyg Assoc J 51:601

72. Jensen B, Wolkoff P (1996) VOCbase-Odor thresholds, mucous membrane irritation thresholds and physico-chemical parameters of volatile organic compounds, version 2.1. National Institute of Health, Denmark

73. Brauer L (1998) Gefahrstoffsensorik. Ecomed, Landsberg

74. Behrendt H and Lemmen C (2002) In: Mücke (ed) Schimmelpilze im Wohnbereich. Gräbner, Bamberg, p 85

75. Pasanen A-L, Korpi A, Kasanen J-P, Pasanen P (1999) In: Johanning E (ed) Bio-aerosols, fungi and mycotoxins: effects, assessment, prevention and control. Boyd, Albany, NY, pp 60–65

76. Kreja L, Seidel HJ (2001) Umweltmed Forsch Prax 6:159

77. Schuchardt S, Kruse H, Wassermann O (2001) Schriftenr InstToxikol Universitätsklinikum Kiel 46:107

78. Pope AM, Patterson R, Burge H (1993) Indoor allergens. National Academy Press, Washington, DC, p 108

79. Gravesen S, Frisvad JC, Samson RA (1994) Microfungi. High Tech PrePress A/S, Copenhagen

80. Hintakka E-L (1998) Indoor Air Suppl 4:66

81. Peat JK, Dickerson J, Li J (1998) Allergy 53:120

The Handbook of Environmental Chemistry Vol. 4, Part F (2004): 179–217
DOI 10.1007/b94835
© Springer-Verlag Berlin Heidelberg 2004

Sensory Evaluation of Indoor Air Pollution Sources

Philomena M. Bluyssen (✉)

Department of Healthy Buildings and Systems, TNO Building and Construction
Research, P.O. Box 49, 2600 AA Delft, The Netherlands
p.bluyssen@bouw.tno.nl

Abstract The basic biological principles of the perception mechanisms for odour and irritants are fairly well understood. Much more uncertain is how these basic processes relate to the more complex psychological responses of odorant/irritant stimulation. Techniques to evaluate air quality with humans are based on measurable attributes such as detection, intensity and quality. The indoor environment comprises thousands of chemical compounds in low concentrations, of which not all can be measured and interpreted by currently available equipment. The nose can detect very low concentrations (parts-per-trillion range) and interpret all at the same time. Besides tobacco smoke, if smoking is allowed, the major indoor air sources comprise furnishings and ventilation systems. Through emission testing of products in laboratory situations, prediction of indoor air qualities in real environments should become possible. However, as long as no unambiguous unit as an indicator for perceived air quality exists, dose-response relations are difficult and labelling even more so.

Keywords Indoor air · Pollution sources · Sensory evaluation

Abbreviations

ECA	European Concerted Action
HVAC	Heating Ventilation and Air Conditioning
IAQ	Indoor Air Quality
IPF	Individual performance factor of panel member (%)
LUR	Lifetime unit risk
PAP	Perceived air pollution
TVOC	Total volatile organic compound
VOC	Volatile organic compound

1
Introduction

From the occupant's point of view, the ideal situation is an indoor environment that satisfies all occupants (i.e. they have no complaints) and does not unnecessarily increase the risk or severity of illness or injury. Both the satisfaction of people (comfort) and health status are influenced by general well-being, mental drive, job satisfaction, technical competence, career achievements, home/work interface, relationship with others, personal circumstances, organisational matters, etc, and last but not least by environmental factors, such as

- Indoor air quality (IAQ): comprising odour, indoor air pollution, fresh air supply, etc.
- Thermal comfort: moisture, air velocity, temperature, etc.
- Acoustical quality: noise from outside, indoors, vibrations, etc.
- Visual or lighting quality: view, illuminance, luminance ratios, reflection, etc.
- Aesthetic quality.

These environmental factors greatly depend on the performance of the enclosure, as well as on the interaction between the human being and the enclosure. People are exposed during more than 90% of their life to these factors in enclosed spaces. Human assessment of the environment is basically expressed by human perception of the environmental factors, and the subsequent assessment of this. One of these factors, IAQ, is becoming more and more a key issue in today's health policies.

The objective performance of the environment can be measured in terms of physical quantities (temperature, noise, illuminance, etc.). The human perception and assessment can be expressed by a person with so-called subjective environmental performance indicators, such as control of environment or specific items (ventilation, noise, light, etc.), acceptability of the environment or a specific item (air quality, thermal comfort, colour, etc.) and complaints or symptoms related to the environment (irritating eyes, skin, headaches, etc.).

The relationship between objective measurement and human assessment is not known for all physical parameters. Mature models for separate subjective issues exist (e.g. thermal comfort [1] and noise) but are not available for all. For example, no consensus model for air quality exists. The reasons are as follows:

- Sensory assessment: The principles behind the sensory evaluation of smell are still under investigation (see Sect. 2).
- Measurement of air quality: The indoor environment comprises thousands of chemical compounds in low concentrations, of which not all can be measured and interpreted by currently available equipment. The nose can detect very low concentrations (parts-per-trillion range) and interpret all at the same time (see Sect. 3).
- Measurement unit: As long as no unambiguous unit as an indicator for perceived air quality exists, dose–response relations are difficult. Total volatile organic compounds (TVOCs) have been used for some time, but the drawback is twofold as they do not represent all pollutants in the air and the effect of single compounds is ignored [2] (see Sect. 4).

The background of the evaluation of air quality using human subjects and the currently available methodologies are presented as well as the sources and chemical compounds measured. Section 2 focuses on the perception mechanisms for odour and irritants. Techniques to evaluate air quality with human panels are described and discussed in Sect. 3. And in Sect. 4 some of the indoor air pollutants are discussed with respect to their odor characteristics and their possible indoor-related sources using the results of several European projects.

2
Perception Mechanisms

2.1
Human Senses

The human perception of IAQ normally involves two human senses, the common chemical sense (somesthesia) and olfaction. Indoor air pollutants that can reach and activate the olfactory epithelium (in the nose) are perceived as odorous. Indoor air pollutants that activate the trigeminal nerve (in mucous

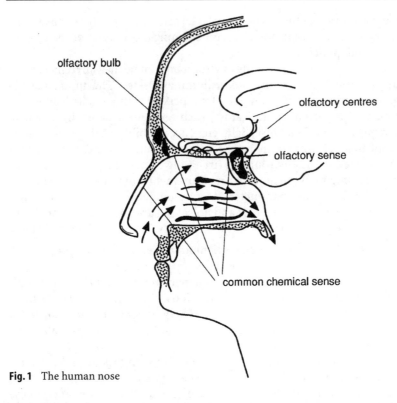

olfactory bulb

olfactory centres

olfactory sense

common chemical sense

Fig. 1 The human nose

membranes of the nasal and mouth cavities, over entire facial and forehead skin) are named irritants. The latter can cause irritation of mucosal membranes in eyes, nose or throat, dryness of eyes, nose and throat, facial skin irritation and dryness of skin.

To be perceived, a molecule is, in general, volatilised from its source, inhaled into the nasal cavity and dissolved in the protective mucous layer (epithelium). The nose comprises two nostrils with smelling organs (Fig. 1). In each nostril one patch of yellowish tissue, the olfactory epithelium, is located in the dome of the nasal cavity. Two types of nerve fibres, the olfactory sense and the trigeminal nerve, whose endings receive and detect volatile molecules, are embedded in this tissue. The trigeminal nerve endings (part of the common chemical sense) are located all over the nasal respiratory lining, not only in the olfactory epithelium. While the olfactory organ is sensitive to the odorant aspect of a chemical, the trigeminal nerve endings are sensitive to the irritant aspect of a chemical in the air. On being stimulated by pollutants, the olfactory nerve endings and the trigeminal nerve endings send signals to the brain, where the signals are integrated and interpreted. The result of this process is called perceived air quality.

The biological principles for receptor activation of odour and irritation are fairly well understood, while the information processes at higher centres of the brain are less clear [3]. How these basic processes relate to the more complex psychological responses to odorant/irritant stimulation, such as perceived air quality, annoyance and symptom reporting, is uncertain.

2.2
Olfaction

The olfactory epithelium comprises between 10 and 25 million olfactory receptor cells, covering a total area of about 4 cm² [4]. Each cell ends in a swollen bulb with a network of hairlike outgrowths (cilia) that extend into the nasal cavity. The fibres from specific epithelium areas are bunched together and end in a glomerule of the olfactory bulb. The olfactory bulb is connected to the brain by a fibre tract. Via the latter, the axons of the bulb are connected with different brain centres, which together are called the olfactory brain.

The sense of smell depends initially upon the interaction between the stimulus and the olfactory epithelium. It is believed that the molecule must be bound by a receptor on the cilia. The "binding" process causes olfactory nerve impulses to travel from the sensory cell to the olfactory lobe of the brain. The brain interprets the incoming signals by associating them with a previous olfactory experience. This is how the nose distinguishes between perceived air qualities [5].

Several theoretical ideas about this process have been developed in the last few decades. These ideas can be divided into two main categories: theories which are based on chemical reactions between the perceived molecules and the tissues in the olfactory cleft, and theories which are based on physical interaction between molecules in the nostrils and the sensitive tissue surface or other physical processes.

The first category comprises a few theories. Kistiakowsky [6] stated that the olfactory response may be set off by a system of reactions that are catalysed by enzymes. A smelling molecule would inhibit the action of one or more of the enzymes. Another theory is that the primary factor determining the odour of a substance might be the overall geometric shape of the molecule. Amoore [7] proposed the stereochemical theory, which provides a mechanism based on a "lock-and-key" principle.

In the second category even more theories are available. For example, Dyson [8] claimed in his Raman shift theory that only substances with shifts of wavelength between 140 and 350 nm, when monochromatic light is shining through them, are detected. Dravniek [9] suggested a hypothetical mechanism based upon a change in coupling between the electron donor/acceptor pair of large molecules which occurs when a substance is absorbed. The altered charge-transfer balance would then be monitored by the appropriate nerve fibres. And Mozell [10] supported the hypothesis that differential sensitivity to substances may be largely a matter of how molecules spread themselves across the olfactory mucosa. The nasal epithelium may act like a gas chromatograph.

The latest research indicates that there are around 1,000 genes that encode 1,000 different odour receptors [11]. Genes provide the template for proteins, the molecules that carry out the functions of the cell. Each type of receptor is expressed in thousands of neurons. Mammalian DNA contains around 100,000 genes, which indicates that 1% of all our genes are devoted to the detection of odours and it shows the significance of this sensory system for the survival and reproduction of most mammalian species. This is in contrast with the human eye. Humans can discriminate among several hundred hues using only three kinds of receptors on the retina. These photoreceptors detect light in different but overlapping regions of the visible spectrum so the brain can compare input from all these types of detectors to identify a colour.

Mammals can detect at least 10,000 odours; consequently, each of the 1,000 different receptors must respond to several odour molecules, and each odour must bind to several receptors. The results shown by Axel [11] suggest that each neuron features only one type of receptor. The problem of distinguishing which receptor was activated by a particular odour is then reduced to the problem of identifying which neurons fired. In all other sensory systems, the brain relies on defined spatial patterns of neurons as well as the position of the ultimate targets of the neurons to define the quality of a sensation. It was found that the olfactory epithelium is divided into four broad regions according to the types of receptors found in each zone, but with a random distribution of receptors within each region, and with no precise spatial pattern of neurons in the epithelium. The results also showed that the glomeruli in the olfactory bulb are differentially sensitive to specific odours. Since the positions of the individual glomeruli are topologically defined, the olfactory bulb provides, therefore, a two-dimensional map that identifies which of the numerous receptors have been activated in the nose. A given odour will activate a characteristic combination of glomeruli in the olfactory bulb; signals from the glomeruli are then transmitted to the olfactory cortex.

Although no specific receptors have been identified yet, it is believed that about 100–300 receptors classes exist. This makes it difficult to predict odour sensations from the chemical structure of an odorant and to establish an objective classification system for odorants.

To summarise the biological process of the olfactory sense:

- An odorant molecule binds to a protein receptor site in the membrane of the receptor cell (at the cilia).
- A receptor is activated by the stimulus (odorous molecule) and in turn activates other proteins which trigger an enzyme cascade (second messenger system), which results in an electric potential.
- Axons of the receptor cells form bundles (glomeruli). Between 30 and 50 such bundles carry olfaction information to the olfactory bulbs.
- Several hundreds of primary olfactory axons converge on a single mitral cell, located on an olfactory synapse of the olfactory bulb.

2.3
Somesthia

It should not be forgotten that the perceived air quality of a substance is the interaction of the perception of the olfactory organ and the common chemical sense (somesthia or trigeminal nerve endings). Cain [12] stated that the difference in the time course between olfactory and the common chemical sensations suggests that different processes of stimulation occur even when stimulation is caused by the same substance. Sensations of the common chemical sense grow during inhalation, but also between inhalations and for relatively long durations, while sensations of the olfactory sense do not build over time. Cain [12] gave three possibilities for explaining the initial buildup of the common chemical sensation. This buildup could be

- An integration of sensation of purely physiological origin, i.e. neural.
- An accumulation of incident stimuli at the neural receptor (majority of trigeminal nerve endings lie somewhat below the surface of the epithelium, so the stimuli may reach the receptors by diffusion; the olfactory receptors are only separated from the atmosphere by a thin layer of mucous).
- Repetitive damage to some structure, such as an epithelial cell which might repetitively secrete endogenous chemicals to serve as the actual stimulus for the common chemical sense (sensation might increase with cumulative damage and might continue after cessation of the stimuli because of low-level inflammation).

Many airborne substances are complex stimuli. They are combinations of many chemicals which can interact at one or several levels before or during the perception: chemical or physical interaction in the gas mixture, interaction of molecules at the receptor surfaces (olfactory and trigeminal systems), peripheral interaction in the nervous system and finally interaction in the central nervous system. Therefore, effects such as masking, neutralisation and counteraction are not surprising in gas mixtures [13].

Only few volatile organic compounds (VOCs) lack the potential to cause irritation. Biological assays to learn the potency of irritants are numerous, for example, the recording of the negative mucosal potential from the nasal epithelium or the recording of cortical evoked potentials, measurement of neural mediators or modulators, measurement of the products of inflammation, and measurement of reflexes [14].

The two types of sensations, trigeminal and olfactory, can be separated by making use of anosmics (persons who lack the sense of smell) as compared with persons with normal olfaction and nasal irritation. To determine the relevant physiochemical determinants of potency, these two groups of persons were exposed to single chemicals [14]. The study showed that a general relationship between odour and irritation thresholds might be present. It was found that the threshold level (odour/irritation) decreases as the length of the carbon chain increases.

2.4
Predicted Versus "Real" Perception

In 1998 two units, olf and decipol, were introduced to quantify sensory source emissions and perceived air quality [15]. This theory is based on the assumption that the pollutants in buildings all have the same relation between exposure and response after one factor normalisation based on human bioeffluents. Emission rates are measured in olf, where 1 olf is defined as the emission rate causing the same level of dissatisfaction as bioeffluents from one seated person at any airflow. Concentration or "perceived air quality" is measured in decipol. One decipol is defined as the concentration of pollution causing the same level of dissatisfaction as emissions from a standard person diluted by a clean airflow of 10 l/s. In this context perceived air quality is the dissatisfaction with or the acceptability of IAQ.

Besides the questionable use of the term perceived air quality, which can involve many parameters other than dissatisfaction and acceptability (e.g. odour intensity, stuffiness, perceived dryness, degree of unpleasantness) [3], the main item that is discussed with this method is the assumption that all pollutants have the same relation between exposure and response, i.e. that the calculated olf values from separate sources can be simply added.

According to Stevens' law the perceived odour intensity of a single compound increases as a power function of concentration [16]:

$$R = C (S - S_o)^n,$$

where R is the perceived odour intensity, S is the stimulus concentration and C, S_o and n are constants.

The perceived intensity of a mixture of two compounds may in theory be as strong as the sum of the perceived intensities of the unmixed compounds (complete addition), more intense than the sum of its compounds (hyper-addition), or less intense than the sum of its compounds (hypo-addition). There are three kinds of hypo-addition:

- Partial addition: the mixture is perceived to be more intense than the stronger compound perceived alone.
- Compromise addition: the mixture is perceived to be more intense than one compound perceived alone, but less intense than the other.
- Compensation addition: the mixture is perceived to be weaker than both the stronger and the weaker compound.

According to Berglund [13], stimulation is proportional to the number of molecules, as long as just one type of molecule is present. The odour interaction for mixtures of constituent odorants is governed by a strongly attenuating function, hypo-addition [17]. The concentration of numerous compounds may be less important to the perceived air quality than the addition or subtraction of a few specific compounds to the gas mixture [18].

Reviewing previous addition studies of perceived air quality (in decipol), it was observed that for the majority of the comparisons between predicted (by using the addition assumption) and measured pollution loads of combinations of sources, the predicted pollution loads are frequently higher than the measured pollution loads [19]; thus, implying hypo-addition as well. The same was found for the perceived air qualities.

Olfactory adaptation, which is similar to visual adaptation to light, makes it even more difficult to predict or model perceived odour intensity or perceived air quality. With continuous exposure, the perceived odour intensity will decrease with time and the odour threshold will increase with time. Recovery or readaptation will occur within less than 1 min after removal from the odour [4].

In conclusion, modelling of perceived quality or intensity of indoor air is not possible yet, based on single compounds, although some predictions have been made with mixtures of several compounds. Indoor air comprises thousands of compounds, of which some are odorous, some are not, and others are irritants. Besides the combined odours and irritant effects of the thousands of compounds, the qualitative character is even more complex. The "sensory" print is, in general, different from the "chemical" print.

3
Sensory Evaluation Techniques

3.1
Attributes

Sensory evaluation of air quality comprises the use of human subjects as measuring instruments. The attributes that can be measured in this way are the same as for all other sensory modalities:

- Detection (the limit value for absolute detection).
- Intensity (odour intensity, sensory irritation intensity).
- Quality (value judgement, such as hedonic tone or acceptability).

In a discrimination evaluation a subject is asked mainly to compare an air sample with another and to express this comparison as "greater than, smaller than or equal to", depending on the attribute evaluated (pleasantness, strength, etc.). For this type of evaluation several techniques are available [3].

3.1.1
Detection

The classical threshold theory assumes the existence of a momentary absolute sensory threshold. However, in real life, there is no fixed odour or irritation threshold of absolute detection for a particular individual or a particular pollutant but rather a gradual transition from total absence to definitely confirmed sensory detection [20]. Therefore, in the theory of signal detectability [21], the same repeated signal is assumed to have a defined distribution, and thus each sensory evaluation by a subject is executed on a probability basis. Berglund and Lindvall [22] have used this signal detection approach to test a few single compounds and in a few building investigations.

In the classical methods the threshold level is defined as the level at which 50% of a given population will detect the odour. One of these methods is the threshold method, which is standardised in many countries for the evaluation of outdoor air [23]. In this threshold method an air sample is diluted stepwise (for each step, by a factor of 2) with clean (odour-free) air to determine the dilution at which 50% of a panel of eight persons can no longer distinguish the diluted air from odour-free air. This number of dilutions, expressed in odour units per cubic metre of air of 20 °C, is the numerical value for the odour concentration of the original air sample. Some measurements using the classical threshold level method, have been made on indoor air, ventilation systems and building materials [22, 24].

The absolute detection threshold varies widely with chemical substances, as is shown by the large spread in odour thresholds for single compounds reported in the literature [25]. This is caused, among others, by the procedure used, the purity of the chemical substance, the equipment applied and the sample of subjects.

Recognition threshold values (the concentration at which a certain chemical is recognised) are usually measured in the same way as detection levels. Both use either the method of limits or the method of constant stimuli [3].

In the method of limits, the chemical substance is presented in alternating ascending and descending series, starting at different points to avoid having the subject fall into a routine. The subject is asked to report whether the sample can be detected or not. The method of constant stimulus is based on the assumption that the momentary individual threshold value varies from time to time and that this variation has a normal distribution. The chemical substance is usually presented in a random selection of concentrations. For both methods, no training is required, although subjects may be selected on the basis of their sensitivity to the chemical substances tested.

3.1.2
Intensity

The intensity of odours or irritants can be obtained by several methods: equal-intensity matching, magnitude estimation or direct scaling methods [3]. The latter is the most common in IAQ studies and uses, for example, visual, semantic scales (e.g. no odour, weak odour, moderate odour, strong odour, very strong odour, overpowering odour). With equal-intensity matching the subject matches the intensity of, for example, two different odorants. Magnitude estimation techniques generate magnitude estimates of intensity resulting from direct numerical estimations by subjects. The perceived intensity of an odour is established by rating the intensity of that odour on a magnitude scale, using reference odours or not. The ASTM technique [26] uses, for example, samples of 1-butanol vapour presented at varying concentrations and the Master Scale unit method [22] uses five concentrations of pyridine which are jointly measured with indoor air samples.

The assessment of decipol levels using trained panels of the air in office buildings (European Audit project [27]) is an example of magnitude estimation with memory references. The same method, but instead having the references (with numerical values) nearby to compare, is an example of magnitude estimation with several references. This method was applied in several European projects (European Audit project [27], Database project [28], MATHIS [29], AIRLESS [30]) and is presented in Sect. 3.2.

3.1.3
Quality

A value judgement of IAQ can be given in several ways. One can make a classification (e.g. yes/no), such as ASHRAE 62-1989 [31] uses (is the air acceptable or not), resulting in a percentage of dissatisfied, or one can use a list of descriptors to describe a chemical substance. The latter is mainly used in the food and perfume industry, from which many classification systems of odours have been developed.

For the evaluation of the acceptability of an air sample (percentage of dissatisfied persons), several methods have been applied. Besides the yes/no classification (acceptable or not acceptable), the continuous acceptability scale [32] is used. The middle of the scale is indicated as the transition between just acceptable and just not acceptable. With both methods, however, large panels (up to 100 persons, depending on the statistical relevance required) of untrained persons are required.

Two units, olf and decipol, were introduced to quantify sensory source emissions and perceived air quality [23]. With these units the so-called decipol method was developed. The decipol method comprises a panel of ten or more persons who are trained to evaluate the perceived air quality in decipol or an untrained panel of at least 50 persons [33]. A method to train a panel to evaluate perceived air quality in decipol has been developed [34, 35]. The latest research indicates, however, that this method does not evaluate the acceptability but the intensity of the air sample (for a description see Sect. 3.2).

3.2
Trained Panel Method

In the context of the European framework for research and technology development, some projects were launched since 1991 aiming at identifying the main causes for indoor air pollution and defining methodologies for this: the European Audit project [27], the Database project [28], MATHIS [29] and AIRLESS [30]. The use of human perception was central in these projects and therefore the trained panel method applied is described in this section. It uses a reference gas, a scale, special equipment and selection and training procedures [36].

3.2.1
Reference Gas and Scale

When a panel has to be trained to evaluate perceived air pollution (PAP), a reference is required. A reference gas that is easy to measure and to produce is 2-propanone [34]. The production of this reference source is based on passive evaporation and the gas is introduced to the human nose by a constant airflow coming out of the so-called PAP meter (formerly named decipolmeter) (see Sect. 3.2.2).

The linear relation between the 2-propanone concentration in air ($C_{2\text{-propanone}}$) and the PAP value is used to set a scale from 1 to 20 [34]:

$$PAP \text{ value} = 0.84 + 0.22 \times C_{2\text{-propanone}} \text{ (ppm)}. \qquad (1)$$

Five different 2-propanone concentrations generated by five PAP meters can be used as the milestones for the training. These milestones have the following concentrations of 2-propanone in the top of the cone of the PAP meter:

- 0 ppm: assigned value 1 (no odour).
- 5 ppm: assigned value 2.
- 19 ppm: assigned value 5.
- 42 ppm: assigned value 10.
- 87 ppm: assigned value 20.

3.2.2
Equipment and Reference Gas Production

The equipment required to select and train a panel of persons comprises about 15 PAP meters, equipment for the production of 2-propanone (reference gas) and an air-conditioned room.

The PAP meter consists of a 3-l jar made of glass covered with a plastic cap, a fan and a diffuser (Fig. 2). The cap has two holes; in one of them the fan is placed to suck the air through the jar. On top of the fan, a cone diffuses the exhausted air. The angle of the cone was chosen to be 8°, to avoid mixing with room air. The diameter of the top of this cone was chosen to be 8 cm, convenient to situate the nose in the middle. The cone is made of glass, supported by a stainless steel stand. The small fan was selected to produce at least 0.9 l/s, several times higher than the highest airflow during inhalation. The person therefore inhales exclusively air from the jar, undiluted by room air.

The 2-propanone gas can be generated in the PAP meter by means of passive evaporation. By placing one or more 30-ml glass bottles filled with 10 ml

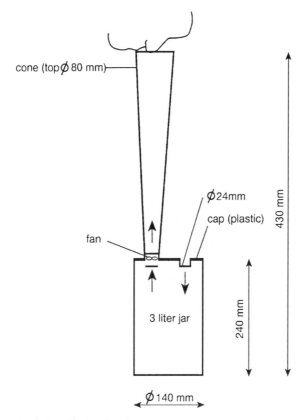

Fig. 2 The perceived air pollution (*PAP*) meter

2-propanone and making different holes in the caps of these bottles, different 2-propanone concentrations can be established.

The steady-state concentration of 2-propanone in the top of the diffuser depends on the level of the liquid in the small bottles (which is standardised at 10 ml), the location of the small bottles in the jar of the PAP meter, the ambient temperature and the variation of this temperature (standardised at 22 °C), and the size of the holes in the caps of the small bottles through which the 2-propanone diffuses.

The time it takes to reach a steady-state concentration depends on several factors: the time and temperature of 2-propanone before it is put in the small bottles, movement of bottles before they are placed in the jar, the transportation of the bottles and the time before the overcaps are removed from the bottles. The following strategy is therefore recommended. Fill the bottles at least 1 h before the test with 2-propanone (that has been conditioned at 22 °C the day before), place the bottles at the correct position in the jar and leave the overcaps off. Activate the fan (6 V required). After 30 min a steady-state level with less than 3% variation should be reached.

Additional recommendations are

- Keep the environmental temperature as constant as possible
  (less than 0.5 °C variation).
- Keep the small bottles at the same location in the jars.
- Keep the combination of jar, fan, cone and small bottles the same if one
  wants to create the same steady-state level again.
- Keep the fan voltage at 6 V.

If only one small bottle, placed on the left side of the two openings in the cap
(Fig. 2), is used to establish a certain concentration, the relation between the
diameter of the hole and the 2-propanone concentration in the top of the cone
is approximately

$$C_{2\text{-propanone}} \text{ (ppm)} = 3 \times \text{diameter (mm)}, \tag{2}$$

where the diameter is 8 mm or less and the standard deviation of 2-propanone
is about 3%.

Combinations of several bottles give, in general, a higher concentration than
the addition of the contents of the individual bottles give. Furthermore, if a
combination of bottles or bottles with different holes are placed in another PAP
meter, concentrations can vary slightly. Once the milestones are calibrated, it is
therefore important to keep the same PAP meters with the same bottles at the
same location in the jar.

The unknown levels to be used for training should vary from 1 to 20, for
example, 1, 1.5, 2, 2.5, 3, 3.5, 4, 4.5, 5, 6, 7, 8, 9, 10, 11, 12, 13, 14, 15, 16, 17, 18, 19
and 20. Each level can be established by putting 1–4 bottles with a specific
diameter hole, filled with 10 ml 2-propanone, in the PAP meter. The location of
the bottles should be noted and they should always be at the same location. For
different numbers of bottles, the recommended locations are shown in Fig. 3.
The unknown concentrations used during the training should, if possible, be
measured every day before the start of the training to determine exactly the
concentration that occurs with the PAP meter used that day. The concentration
might vary although the same bottles at the same locations are applied.

The position of the small bottles in the PAP meter is of great importance. To
make it easier to reproduce the positions of the small bottles, a scale glued at the
bottom of the PAP meter is a possibility. Furthermore, a sign on the cone of the
PAP meter and the 3-l jar of the PAP meter is handy to keep them together.

The training of the panel should take place in a well-ventilated, temperature-
controlled space with low-emitting and low-ab(ad)sorbing/desorbing materials.

A critical point in the use of the PAP meter, as an instrument to produce
different 2-propanone levels to train a panel to evaluate perceived air quality,
is the establishment of the low 2-propanone values, i.e. values below 1 on the
scale. A zero level can not be established by the PAP meter as such. The PAP
meter without any 2-propanone results in a PAP level of about 0.8. To prevent
this deficiency from increasing, it is therefore of utmost importance that the
training takes place in a room with a very low background level.

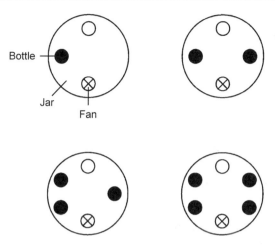

Fig. 3 Recommended locations of small bottles in the PAP meter

The space where the sensory panel is trained has to fulfil certain criteria. A space which is temperature-controlled, has 100% outdoor ventilation, a filtration unit (e.g. active carbon), a Teflon layer on the walls, the floor and the ceiling, and displacement ventilation (from floor to ceiling) or a local exhaust is preferable. A space which is an empty room (no smoking), has walls, a floor and a ceiling that could be covered with a Teflon layer or cleaned with a non-smelling agent, has a mechanical air supply with filtered air and has mixing ventilation with a certain minimum ventilation rate is acceptable.

The minimum criterion that a space where the sensory panel is trained should follow is a background level of 2-propanone expressed in a maximum allowable concentration. This maximum allowable concentration is 1 ppm.

3.2.3
Selection and Training Procedure

A panel of 12–15 subjects should be selected for the training. The subjects should be selected from a group of at least 50 applicants of ages ranging from 18- to approximately 35-years old. There is no restriction on distribution of gender or smoking habits.

Each of the applicants should participate in a selection test. The subjects should be asked before the selection test to abstain from smoking and drinking coffee for at least 1 h before the test. Also, they should be asked not to use perfume, strong smelling deodorants or make-up, and not to eat garlic or other spicy food the day before the test and on the day of the test.

In the selection test, the applicants will one by one be given a short introduction in how to use the milestones and in how to put their nose in the cone of the PAP meters. They will then be asked to assess eight different concentrations of 2-propanone using the milestones as the reference. The applicants will be instructed to have at least two inhalations of unpolluted air in between each exposure to a 2-propanone concentration. The question that is asked to the applicants is the following: "How strong is the air that you perceive? Give a number on a scale from 1 to 20, but always refer this number to the numbers on the milestones 1, 2, 5, 10 and 20. One is equal to no smell (you perceive nothing), 20 is equal to extremely strong smell."

During the test, the applicants are allowed to go back and forth between the eight different unknown concentrations and the milestones as often as they need. Five of the concentrations should be evenly distributed in the range 1–10, while the last three concentrations should be in the range 10–20. For each person the sum of the numerical errors (differences between the voted and the correct values) in the eight assessments is calculated. The 12–15 subjects with the lowest sum of errors are then selected.

The 12–15 subjects will be trained for 3–5 days in smaller groups of three or four persons. Each day they will receive approximately 1 h of intensive training. In the first 2 days of the training the panel will be trained to assess the PAP of concentrations of 2-propanone unknown to them by making a comparison with the milestones. On the third to the fifth day, training will comprise 2-propanone concentrations and other sources of pollution.

Before the training starts, and if necessary this will also be stressed during the training, the panel members will be asked again to abstain from smoking and drinking coffee for at least 1 h before the test. Also, they are asked not to use perfume, strong smelling deodorants or make-up, and not to eat garlic or other spicy food the day before the test and on the day of the test.

On the first day of the training, the subjects will receive instructions about the training procedure and the experiments. During the refreshment time (or waiting time) the panel members are placed in a well-ventilated room where they can talk together but they are not allowed to talk about the experiments. The instruction they received during the selection will be repeated and it will be emphasised that they should always take two inhalations of unpolluted air before they are exposed to another source to prevent them getting used to the smell.

The panel members will be instructed in how to use the milestones and the scale correctly. They will be instructed to rate the intensity of the test concentration with a number from 0 to 20 by making a comparison with the intensity of the milestones. The panel members are allowed to go back and forth between the milestones and the unknown concentration or source, but they will be instructed to have at least two inhalations of unpolluted air between each exposure to avoid adaptation. After the evaluation of each unknown concentration of 2-propanone the panel member will be given the correct answer and the performance of the panel member will be discussed with the experiment leader. The panel members will write their vote on a form if it is possible to follow their performance during the training.

The panel members will be exposed to 6–12 2-propanone concentrations during each training session. From the second day of training the subjects will furthermore be trained to assess air polluted with samples of building materials or other sources. Since the pollutants have a different character than 2-propanone it is of great importance that the subjects understand that they are exposed to the intensity by comparing the intensity of the milestones. For the assessments with air samples of sources other than 2-propanone, the perceptions cannot be compared with any expected result but the assessments can be discussed with each panel member separately.

On the third day of the training the panel members are exposed to a performance test (see Sect. 3.2.4.2). The panel members assess the concentrations as during the previous training except that no feedback is given on the assessment deviations. If the votes do not meet the requirements, the panel member does not qualify and should be trained at least one more day before taking the exam once more. Another option is to exclude the panel member from the panel.

On each day of experiments, the panel members will be retrained for approximately 15–20 min per group of 3 or 4 persons. During this training the panel members will be exposed to two or three different concentrations of 2-propanone and two different materials, which they will receive feedback on. Also, on each day of experiments, the panel members will be exposed to six different concentrations of 2-propanone corresponding to the values 1, 3, 7, 12, 16 and 19. The concentrations of 2-propanone should be measured just before the sensory assessments of each group of panel members. These exposures make it possible to compare different sensory panels and to calculate performance factors (see Sect. 3.2.4).

The panel members will be placed in a well-ventilated waiting room. During each round of assessments, the subjects will one by one assess the intensity of the PAP of the air sample (from a material in a PAP meter, in a walk-in climate chamber, air from a ventilation system, etc.) by making a comparison with the intensity of the milestones. The panel members will be allowed to go back and forth between the milestones and the polluted air sample. The subjects will write down their assessment on a voting sheet, which they will hand to the experimental leader before making the next assessment.

The time between assessments for each panel member should not be less than 3 min. An experienced panel member can assess an air sample within 30–45 s. With a panel of 12 subjects the time between assessments for a panel member will be approximately 9 min, and for a group of four panel members 3 min.

3.2.4
Performance

The training level can be determined by using the given votes compared with correct votes for the 2-propanone levels and by using the repeated votes and/or the standard deviation on the panel vote for the unknown sources.

3.2.4.1
Performance with 2-Propanone

Every training day a panel evaluates a certain number of samples for their unknown 2-propanone levels. A linear regression of all given votes versus the correct votes for each panel member or the whole panel (the ideal relation, i.e. a perfectly trained panel member, is then voted equals correct), can then be determined. On the basis of these lines, each individual panel member can be instructed how to adjust his or her votes.

The difference between the correct and the voted level shows how the vote lies towards the line voted equals correct. A panel member can have an almost perfect relation between the voted and the correct level but still have large differences between the correct and the voted level. It is therefore important to take the relative difference between the correct and the voted level into consideration. This can be defined by the so-called performance index:

$$PF = (\text{voted} - \text{correct}) \times 100/\text{correct} \ (\%), \tag{3}$$

where PF is the performance index (as a percentage), voted is the voted level and correct is the correct level.

To determine the training level per day, the mean of all performance indices of all given votes that day can be calculated, together with the standard deviation of the performance index. A panel member with a performance index of 9% with a standard deviation of 60% performs worse than a panel member with a mean performance index of –20% and a standard deviation of 10%. The second panel member is more consistent in his or her votes than the first one.

Each day the panel members can now be ranked according to their best performance by adding the square root of the quadratic performance index to the standard deviation. The best panel member is the one with the lowest result.

Another way to determine the training level of a panel, is to calculate the standard deviation of a single vote given to the unknown levels of 2-propanone. The mean standard deviation for all evaluated levels per day presents a training level. This statement assumes, however, that the standard deviation is independent of the evaluated level.

3.2.4.2
Individual Panel Member Exam

For the individual panel exam, each individual panel member must evaluate six 2-propanone levels with the values 1, 3, 7, 12, 16 and 19, in a random order. The votes have to meet certain requirements. If the votes do not meet these requirements, the panel member does not qualify and should be trained for at least one more day before he or she is allowed to take the exam once more. Another option is to exclude the panel member from the panel.

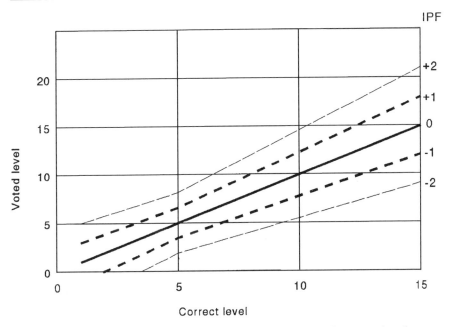

Fig. 4 Required precision of an individual panel member's vote when assessing the perceived air quality of 2-propanone concentrations. For the individual exam the following counts are accepted: votes in the area between the *fat line* and the *fat dashed line*. One in four votes is accepted in the area between the fat dashed line and the *thin dashed line*

The requirements for the individual exam are shown in Fig. 4. In this figure the individual performance factor is shown for each combination of voted and correct level [37, 38].

3.2.4.3
Individual Performance Factor

For comparison between panel members of different panels the individual performance with 2-propanone concentrations can be described by the individual performance factor (IPF). The IPF is defined as

$$PF = \text{voted error/allowed error} = (\text{voted} - \text{correct})/(A \times \text{correct} + B), \quad (4)$$

where the voted error is the voted level minus the correct level, the allowed error is the allowed difference between the voted level and the correct level, A is the tangent of the angle difference between lines and B is the intersection with the y-axis. For PAP<5, $A=-3/28$ and $B=59/28$, for PAP>=5, $A=4/28$ and $B=24/28$. The values for A and B originate from experience with trained panels in the EC Audit project and are related to perceived air quality evaluations in decipol [38].

For this performance method the ideal vote (voted equals correct) is taken as index 0. Besides that allowed maximum and minimum errors (both level-dependent) are defined as index +1 and –1. With this approach the index should theoretically be independent of the level chosen (if the error limits are chosen correctly), and should result in information about the voted error related to the allowed error (0 is perfect; less than |1| is allowed; more than |1| is bad).

The mean value of the IPF and the standard deviation of the IPF give an indication of the quality of the panel member related to 2-propanone concentrations. If the allowed errors change, A and B in the formula change as well.

3.2.4.4
Panel Performance Factor

The same approach is possible for the whole panel. The mean IPF value of the performance for the whole panel is called the panel performance factor. The mean value of the panel performance factor and the standard deviation give an indication of the quality of the whole panel related to 2-propanone concentrations.

3.2.4.5
Performance with Other Sources

The performance of the whole panel for sources other than 2-propanone can be shown by the standard error of the mean votes given to the perceived air quality caused by sources other than 2-propanone, and by the reproducibility for replicas (standard deviation of the replicas).

Several judgements of the same pollution source provide information on the reproducibility of a panel. The standard deviation around the mean of two or more replicas of a source divided by the mean vote of that source determines the reproducibility.

3.3
Human Nose Versus Electronic Nose

The development of instruments, an artificial nose or an electronic nose, that can evaluate the air quality as the human nose does is an ongoing activity. Many attempts have been made, some successful for the purpose they are designed for, others not. The reason is not only related to the still incomplete knowledge of the perception mechanism (information processes in the brain), but also to the fact that the nose is able to detect very low concentrations.

The lowest odour detection level that could be found in the literature is presented in Table 1 for a number of compounds that are emitted by the human body [34]. From this table it follows that the human nose is able to detect certain compounds at the parts-per-trillion level.

Table 1 Odour detection levels for some compounds emitted by the human body [34]

Compound	Molecular weight	Structure	Odour detection level	
			$\mu g/m^3$	ppb[a]
Acetaldehyde	44	CH_3CHO	0.2	0.111
Benzaldehyde	106	C_6H_5CHO	0.8	0.185
Butyric acid	88	CH_3CH_2CH- $(OCCH_3)CO_2C_2H_5$	1	0.278
Coumarin	146	$C_9H_6O_2$	0.007	0.0012
Dimethyl sulfide	62	$(CH_3)_2S$	2.5	0.986
Dimethyl disulfide	94	CH_3SSCH_3	0.1	0.026
n-Decanal	156	$CH_3(CH_2)_8CHO$	0.25	0.039
Ethanethiol or ethyl mercaptan	62	C_2H_5SH	0.1	0.039
Hydrogen sulfide (inorganic)	34	H_2S	0.7	0.503
Methyl mercaptan or methanethiol	48	CH_3SH	0.04	0.020
Phenylacetic acid	136	$(C_6H_5CH_2CO)_2O$	0.03	0.0054

[a] ppb = 24.45 × ($\mu g/m^3$)/molecular weight.

Furthermore, the results of Cain and Cometto-Muniz [14] indicate that complex chemical environments may enable chemosensory and particularly irritative detection when single VOCs lie far below their individual thresholds. This means that in gas mixtures the nose may detect even far below single thresholds, i.e. below the parts-per-trillion to parts-per-billion range.

It should be well understood that the use of human subjects to evaluate perceived air quality, the so-called sensory evaluation of air, is only one way of measuring the air quality. Compounds such as carbon monoxide cannot be smelled by a human being and can nevertheless be health-threatening. Those compounds should therefore be measured in another way.

The methods or instruments available to measure indoor air compounds can be divided into two groups [39]:

- Those that require an extraction step before making a physical or chemical measurements (e.g. chromatography).
- Those that make a direct physical measurement of some property of the sample (e.g. nondispersive IR spectrometry).

Chromatography is a separation technique in which an inert gas or liquid (mobile phase) flows at a constant rate in one direction through the stationary phase, a solid with a large surface-to-volume ratio or a high boiling liquid on a solid support. The sample may be a gas or a liquid, but it must be soluble in the mobile phase. Gas chromatography is used for separation of volatile, relatively nonpolar materials or members of homologous series; liquid chromatography is used for separation of particularly those materials with low volatility and labile or instable compounds; and thin layer and column chromatography are used for separation of inorganic or organic materials, and low molecular weight species up to high-chain-length polymers.

Spectrometry or photometric methods make use of discrete energy levels of molecules and the emission or absorption of radiation which usually accompanies changes by a molecule from one energy level to another. They are generally based on the measurement of transmittance or absorbance of a solution of an absorbing salt, compound or reaction product of the substance to be determined. They include absorption spectroscopy, emission spectroscopy, laser spectroscopy, photoacoustic techniques and X-ray analysis. In photometry it is necessary to decide upon the spectral levels to be used in the determination. In general, it is desirable to use a filter or monochromator setting such that the isolated spectral portion is in the region of the absorption maximum. A monochromator is a device or instrument that, with an appropriate energy source, may be used to provide a continuous calibrated series of electromagnetic energy bands of determinable wavelength or frequency range.

Mass spectrometry and flame ionisation can be placed under the category ionisation methods. In mass spectrometry a substance is made to form ions and then the ions are sorted by mass in electric or magnetic fields. Positive ions are produced in the ion source by electron bombardment or an electric discharge.

A flame ionisation detector makes use of the principle that very few ions are present in the flame produced by burning pure hydrogen or hydrogen diluted with an inert gas. The introduction of mere traces of organic matter into such a flame produces a large amount of ionisation. The response of the detector is roughly proportional to the carbon content of the solute. The response to most organic compounds on a molar basis increases with molecular weight.

Chemical sensors for gas molecules may, in principle, monitor physisorption, chemisorption, surface defects, grain boundaries or bulk defect reactions [40]. Several chemical sensors are available: mass-sensitive sensors, conducting polymers and semiconductors. Mass-sensitive sensors include quartz resonators, piezoelectric sensors or surface acoustic wave sensors [41–43]. The basis is a quartz resonator coated with a sensing membrane which works as a chemical sensor.

With conducting polymers, a wide range of aromatic and heteroaromatic monomers undergo electrochemical oxidation to yield adherent films of conducting polymer under suitable conditions [44]. The conductivity of the polymer film is altered on exposure to different gases.

The principle of semiconductor sensors is based on the change of the electrical characteristics of the semiconductor when the gas to be measured is absorbed. The change of the number of free load carriers or the change of polarisation of the bounded load carriers is then measured [43].

Several commercial instruments are available. Some comprise conducting polymers, others tin oxide gas sensors (thick or film devices) or metal oxide semiconductors, and combinations. However, none of them can evaluate IAQ as the nose does.

4
Indoor Air Pollutants and Their Sources

4.1
Indoor Air Pollutants

The main groups of pollutants found in indoor air are chemical and biological pollutants. Among the chemical group one can distinguish gases and vapours (inorganic and organic) and particulate matter. And among the biological group belong microorganisms: mould, fungi, pollens, mites, spores, allergens, bacteria, airborne infections, droplet nuclei, house dust and animal dander. The main groups of pollutants found in indoor air are presented in Table 2.

Inorganic gases (NO_x, SO_x) are, in general, not odorous, except for some such as ammonia and sulfur dioxide. The same can be said of biological pollutants, with the exception of some products that are excreted by microorganisms. Particulate matter can only partly reach the nose. On the other hand, virtually all organic vapours stimulate olfaction [44].

Under normal conditions, most nonvolatile chemicals cannot reach the human olfactory epithelium and so are unable to stimulate olfaction. However, such molecules, when presented as aerosols, can reach the sensory tissue and can then stimulate a response. Odorants are typically small, hydrophobic, organic molecules with a mass range of 34–300 Da. Most odorants contain a single polar group. The majority of odorants contain oxygen.

Table 2 Main groups of indoor air pollutants

Groups	Subgroups	
Chemical	Gases and vapours	Inorganic: NO_x, SO_x Organic: volatile organic compounds, CO, formaldehyde
	Particulate matter	Asbestos, respirable particles with a diameter less than 10 μm, particulate matter which is smaller than 10 μm
	Radioactive particles/gases (radon & its daughters)	
Biological	Microorganisms, mould, fungi, pollens, mites, spores, allergens, bacteria, airborne infections, droplet nuclei, house dust, animal dander	

Table 3 Classification of organic indoor pollutants [45]

	Boiling point range (°C)	Sampling methods typically used
Very volatile organic compounds	<0 to 50–100	Batch sampling, adsorption on charcoal
Volatile organic compounds	50–100 to 240–260	Adsorption on Tenax, graphitised carbon black or charcoal
Semivolatile organic compounds	240–260 to 380–400	Adsorption on polyurethane foam or XAD-2[a]
Organic compound associated with particulate matter	>380	Collection on filters

[a] Styrene–divinylbenzene copolymer.

Table 4 Chemical structures of volatile organic compounds most frequently detected indoors and examples [46]

Chemical structure	Examples
Alkanes	n-Hexane, n-decane
Cycloalkanes and alkenes	Cyclohexane, methylcyclohexane
Aromatic hydrocarbons	Benzene, toluene, xylene
Halogenated hydrocarbons	Dichloromethane, trichloroethane
Terpenes	Limonene, α-pinene
Aldehydes	Formaldehyde[a], acetaldehyde[a], hexanal
Ketones	Acetone, methylethylethanol
Alcohols, alkoxyalcohols	Isobutyl alcohol, ethoxyethanol
Esters	Ethylacetate, butylacetate

[a] Not a volatile organic compound.

The World Health Organisation classified organic indoor pollutants in four categories [45] (Table 3). Furthermore, the European Concerted Action (ECA) made a division in the structures of chemicals which are mostly detected indoors (Table 4) [46].

Particles are defined as aerosols (dust) when they are smaller than about 200 μm and larger than 0.01 μm. Smaller particles have the characteristics of a gas, and larger particles are too heavy to stay suspended and will not be inhaled. Inhalable particles which can reach the pharynx have a maximum size of 200 μm, particulate matter which is smaller than 10 μm can reach the larynx and the thorax, and respirable particles can go as far as the alveoli in the lungs (e.g. asbestos fibres) [47]. Particles with a diameter between 2 and 5 μm can precipitate in the alveoli; even smaller particles are exhaled again. Air sample analysis indicates that up to 99% (by count) of particles present in the atmosphere are 1 μm or less in size [48].

Mycotoxins are chemicals manufactured by fungi, some of which are extremely toxic to humans and animals [49]. When moulds make them, they also make synergisers, substances that can enhance the potency of other toxins in the environment. Some of these compounds may not be toxic in themselves but become toxic when combined with other substances.

Fungi also emit VOCs, which are responsible for their odour. More than 500 VOCs have been identified from different fungi. One of the commonly produced VOCs, ethanol, is very volatile and acts as a potent synergiser.

4.2
Indoor Air Sources

The possible sources of indoor air pollution can be categorised into

- Outdoor sources: traffic, industry.
- Occupant-related activities and products: tobacco smoke, equipment (laser printers and other office equipment), consumer products (cleaning, hygienic, personal care products).
- Building materials and furnishings: insulation, plywood, paint, furniture (particle board), floor/wall covering, etc.
- Ventilation systems.

In the European Audit project to optimise indoor quality and energy consumption in office buildings, 56 office buildings in nine European countries were audited during the heating season of 1993–1994 [27]. In this audit, besides normal measurements such as questionnaires and the physical/chemical analysis of air, panels of persons, trained to evaluate the perceived air quality, were used to measure the air quality in preselected spaces of those office buildings as well as the outdoor and supply air. From this investigation it was concluded that the main pollution sources were the materials, furnishing and activities in the offices and the ventilation system in the buildings.

The possible sources for the most important chemical compounds identified in the European Audit project are presented in Table 5 [50]. The most important source of VOCs was materials, especially furnishings. The dominant VOCs detected in the majority of the buildings were solvents used in floor or wall coverings and pressed-wood products (carpets, PVC flooring, floor adhesives, wallpaper, particle board, etc.).

As a follow-up to the European Audit project, the Database project (European database on indoor air pollution sources in buildings) was launched to investigate the emissions (sensory and chemical) of indoor air sources, and in particular the building materials, more closely. This resulted in the first database of indoor air pollution sources [28].

To get more detailed information on the building materials, the European project MATHIS (materials for healthy indoor spaces and more energy efficient buildings) [29] was executed, together with the European project AIRLESS (a European project to optimise air quality and energy consumption of heat-

Table 5 Possible sources of the most prevalent volatile organic compounds found in audited buildings as found in a literature survey [50]

No.	Compound	Sources O[a]	Sources T[b]	Materials E[c]	Materials B[d]	Materials F[e]	Materials C[f]
1	(CF₂)ₙ						x
2	1,1,1-Trichloroethane					x	x
3	C₂Cl₃F₃						x
4	Tetrachloroethylene						x
5	Dichloromethane				x		x
6	Dichlorobenzene						x
7	Butane	x					
8	n-Hexane	x	x	x		x	x
9	Aliphatic C₇H₁₆	x					x
10	n-Heptane	x					x
11	Octane	x	x	x			
12	Aliphatic C₉H₂₀			x		x	x
13	Nonane			x		x	x
14	Decane C₁₀H₂₂			x		x	x
15	Undecane			x		x	x
16	Dodecane						x
17	Tetradecane						x
18	Pentadecane	x					
19	2-Methylbutane	x					
20	2-Methylpentane	x					
21	3-Methylpentane	x					
22	2,4-Dimethylhexane	x				x	
31	Benzene	x	x			x	
32	C₃ alkylbenzenes	x	x		x	x	x
33	m-Xylene	x	x		x	x	x
34	o-Xylene	x	x		x	x	x
35	p-Xylene	x	x		x	x	x
36	Toluene	x	x	x	x	x	x
37	Naphthalene						
38	Phthalate compounds						
39	1-Butanol				x	x	x
40	1-Ethoxy-2-propanol				x		
41	2-Butoxyethanol				x	x	x
42	2-Phenoxyethanol				x	x	x
43	C₅ alcohol				x		x
44	Ethanol					x	x
45	Ethoxy-ethoxy-ethanol						x
46	4-Methyl-2-pentanone		x		x	x	x
47	Acetone						
48	Cyclohexanone						
49	Benzaldehyde			x			x
50	Nonanal			x	x	x	x
51	Decanal			x		x	x
52	Acetic acid butyl ester			x	x	x	x

Table 5 (continued)

No.	Compound	Sources		Materials			
		O^a	T^b	E^c	B^d	F^e	C^f
23	2-Methylhexane	x				x	
24	Nonane/o-xylene					x	
25	Nonane/styrene					x	
26	Dimethylcyclopentane					x	x
27	Methylcyclopentane					x	x
28	Methylcyclohexane	x				x	x
29	Cyclohexane					x	x
30	2-Methyl-1,3-butadiene					x	

No.	Compound	Sources		Materials			
		O^a	T^b	E^c	B^d	F^e	C^f
53	Acetic acid ethyl ester					x	
54	Butoxy-ethoxy-ethylacetate					x	x
55	Acetic acid				x	x	
56	Benzoic acid						x
57	Dodecanoic acid					x	x
58	α-Pinene					x	x
59	l-limolene						x
60	Terpene compounds						x

a Outdoor air.
b Tobacco smoke.
c Office equipment.
d Building materials.
e Furnishings.
f Consumer products.

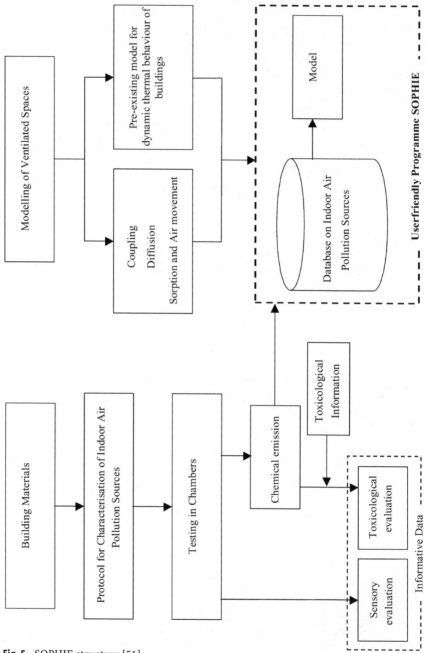

Fig. 5 SOPHIE structure [51]

ing, ventilation and air-conditioning, HVAC, systems) [30], which was focussed on pollution sources from ventilation systems. Both resulted in an extended version of the database named SOPHIE [51].

SOPHIE is the acronym adopted for a database of indoor air pollution sources, including building materials and furnishings and ventilation-system components [51]. It represents the result of the work of a vast network of laboratories in Europe developed under the sponsorship of the European Commission. It aims to document the most important indoor pollution sources and to create a model to establish a link between the strength of the pollution sources and the ventilation rate and its consequences in terms of the IAQ in a given space.

SOPHIE is a tool that can become a reference database and function as a basis for launching more practical or specific databases and labelling frameworks at different levels. This could be by differentiating construction products or by reflecting the diversity of state or national contexts. Its data can be handled and compared with a particular high degree of confidence as its results have been obtained from different laboratories following the same protocols and checked through different processes (i.e. pilot studies of intercalibration at the European level) [52]. The structure of SOPHIE is illustrated in Fig. 5.

It contains information on some HVAC components but the more prevalent information refers to construction materials. A balance of the type of materials tested and of how many tests were performed during the two major campaigns in the development process of SOPHIE is made in Table 6. Those materials were tested according to standard procedures for testing chambers [53] and for chemical analysis [54]. A sensory assessment was also performed. Two product

Table 6 Description of emission sources and related tests investigated for SOPHIE [51]

Type of source	Number of sources	Number of chemical tests			Number of sensory tests		
		3rd day	14th day	30th day	3rd day	14th day	30th day
Flooring	56	53	14	55	57	28	57
Wall	33	26	5	29	14	0	14
Ceiling	2	2	0	2	0	0	0
Construction	14	10	3	14	17	8	17
Other	14	13	2	14	11	6	11
Heating, ventilation and air-conditioning components[a]	16	18	0	0	39	0	0
1994–1997	85	86	24	68	106	32	67
1998–2000	50	46	0	46	32	10	32

[a] Tests not related to time but to air volume.

ages were generally considered for the materials tested: 3 and 30 days. In some cases an intermediate age of 14 days was also considered.

SOPHIE is more than just a static list of materials organised by several different criteria related to their contribution to the IAQ. Once the emission rate of chemicals by the different materials had been determined, the next step was to establish the model to link the concentrations of those chemical substances (pollutants) in a given space with a certain level of ventilation rate and with the dynamics of a certain thermal environment indoors. That is why a major tool is incorporated to enable the linkage of the ventilation conditions with the actual level of the IAQ conditions for a certain type of occupation, including the nature and the extension of materials employed. It is a quite ambitious dynamic model but probably with the merit of establishing a frame with enough generality and broadness to allow for further developments.

The model has certainly many limitations, some related to the status of current knowledge and others due to the lack of appropriate information. It is clear nowadays that sorption/desorption effects play an important role in the actual levels of the concentrations of certain pollutants, depending on the different ambient conditions, but, above all, in the interaction of the material/substance. The fact is that the values for specific coefficients of adsorption or desorption of a particular chemical substance in a given material are generally unknown. Recent studies have been made in order to obtain the information needed on the sorption/desorption coefficients for different coupling material/substances [29]. So far, that work has only been done for a very limited number of cases. Given the hundreds of substances and materials that can be present, that limitation probably represents the major bottleneck to the wide application of the model.

4.3
Ventilation Systems

For ventilation systems a separate study named AIRLESS was performed [30, 55]. Experiments were performed to investigate why, when and how the components of HVAC systems pollute or are the reason for pollution. Different combinations of temperature, relative humidity, airflow and pollution in passing air were investigated. Measurements of perceived air quality, particles, chemical compounds (such as very volatile organic compounds and aldehydes) and biological compounds were selected for each component. The most polluting components of HVAC systems were studied in the laboratory and in the field. The perceived air quality or odour intensity was in most cases measured with a trained sensory panel, according to the protocol developed for the AIRLESS project (Sect. 3.2).

It was concluded that main sources and reasons for pollution in a ventilation system may vary considerably depending on the type of construction and the use and maintenance of the system. In normal comfort ventilation systems the filters and the ducts seem to be the most common sources of pollution, especially odours. If humidifiers and rotating heat exchangers are used, they are also suspected to be remarkable pollution sources especially if not constructed and maintained properly. The pollution load caused by the heating and cooling coils seems to be less notable. And the effect of airflow on the pollution effect of HVAC system components seems to be less important.

4.3.1
Filters

Filters are one of the main sources of sensory pollution in ventilation systems [56]. New filters already seem to influence the odour intensity negatively. The filter material had a significant influence on the starting pollution effect of new filters (Fig. 6). The pollution of new filters decreased after some time of use. When the filters got older, i.e. were in use for some time, the pollution increased again. The reason for pollution after the filter is in use for some time is still unclear. It seems that microorganisms may not be the only pollution source on a filter. Environmental conditions such as airflow (amount of intermittent/continuous flow) and temperature did not have an influence on the pollution effect.

4.3.2
Ducts

The duct material and the manufacturing process had the biggest effect on the perceived air quality [57]. Depending on the machinery used in the manufac-

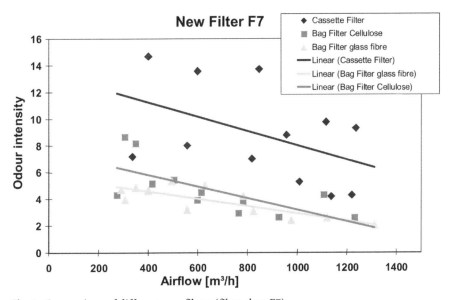

Fig. 6 Comparison of different new filters (filter class F7)

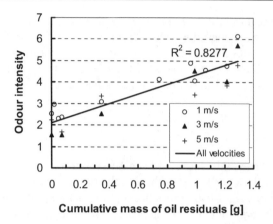

Fig.7 Correlation between odour intensity and the mass of oil residuals in the ducts tested

turing process, new spiral wound ducts, flexible ducts and other components of the ductwork might contain small amounts of processing oil residuals. The oil layer is very thin and invisible, but it emits an annoying odour. Aluminium ducts score the best with respect to odour intensity. Plastic ducts seem a feasible solution.

Oil residuals are the dominating sensory pollution source in new ducts. The sensory assessments showed a clear correlation between the total mass of oil residuals (average surface density times surface area) and the PAP (Fig. 7). Emissions from dust/debris accumulated in the ducts during construction (mostly inorganic substances) seem to be less important. No simple correlation was observed between the amount of accumulated dust and odour emissions; however, the organic dust accumulated during the operation period may produce more severe odour emissions. When dust had accumulated on the inner surface of the ducts, the relative humidity of the air in the ducts had an effect. On the other hand, the relative humidity had virtually no effect on the odour emissions of oil residuals.

The effect of airflow on the odour intensity from ducts was relatively small and is probably insignificant in normal applications. The length of the duct had a significant influence on the odour intensity. The longer the duct the worse is the odour intensity at the end if the ducts are not clean.

4.3.3
Humidifiers

The main reasons of pollution from humidifiers were determined, namely, disinfecting additions, old water in tanks and/or dirty tanks, microbiological growth, wrong use when the humidifier is off (the water stays in the tank too long), and desalinisation and demineralisation devices/agents (if used) [58]. Humidifiers only pollute the air significantly if the humidifier is not used in the

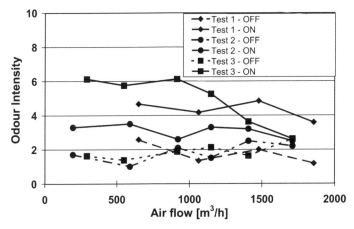

Fig. 8 Odour intensity for a steam humidifier

prescribed way or if it is not properly maintained. The investigations make it clear that periodical cleaning of humidifiers is an absolute must, as is the use of fresh water (Fig. 8). Under normal conditions, it was found for all humidifiers that the airflow has no influence on the odour intensity caused by humidifiers.

A relation was found between the odour intensity and the concentration of bacteria on the inside of the humidifier (Fig. 9). The odour intensity increases with increasing number of bacteria. This was not the case for other locations in an HVAC system. A similar correlation could not be found for fungi.

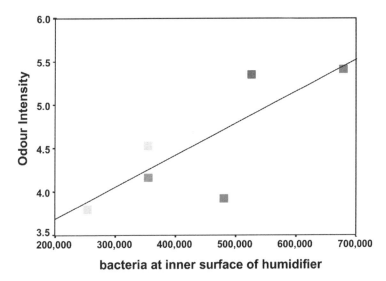

Fig. 9 Bacteria concentration at the inner surface of a humidifier correlated with odour intensity

4.3.4
Rotating Heat Exchangers

Rotating heat exchangers may transfer contaminants from exhaust to supply air in three ways: with entrained air, through possible leakage around the wheel at the separation wall and by adsorption/desorption on the inner surfaces of the exchanger's wheel.

Leakage from exhaust to supply was measured in several units, and was found to be negligible in most cases. Leakage and pollutant transfer can be avoided or at least strongly reduced through the proper installation of the wheel, good maintenance of the gasket, proper installation of a purging sector and by maintaining a positive pressure differential from supply to exhaust duct at wheel level.

Significant amounts of VOCs are transferred when the purging sector is not used well [59, 60]. Even when it is installed well, certain categories of VOCs are easily transferred by a sorption transfer mechanism. Among the VOCs tested, those having the highest boiling point were transferred best. The largest transfer rate in a well-installed unit was found for phenol (30%) (Fig. 10).

4.3.5
Coils

The results showed that heating and cooling coils without condensed water or stagnant water in the pans are components that have small contributions to the overall odour intensity of the air. However, cooling coils with condensed water in the pans are microbial reservoirs and amplification sites that may be major sources of odours to the inlet air.

4.4
Ranking and Labelling

The ultimate goal for emission testing of building products or even built environments is to provide acceptable (healthy and comfortable) IAQ for the occupants. Through emission testing of products in laboratory situations, prediction of IAQs in real environments should become possible. SOPHIE is the first attempt at this prediction and the first attempt to include sensory evaluation and not merely chemical emission testing.

For labelling purposes of building products one further step has to be made. In report 18 of the ECA [61] such an attempt was made, comprising a procedure for testing, evaluating and labelling of flooring materials. In this procedure, a flooring material is tested according to a strict procedure:

- After 24 h of conditioning in a testing laboratory: to protect the panel members screening of emissions for specified compounds; for specified carcinogenic VOCs the lifetime unit risk (LUR) should be less than 10^{-4}; if not no label is given.

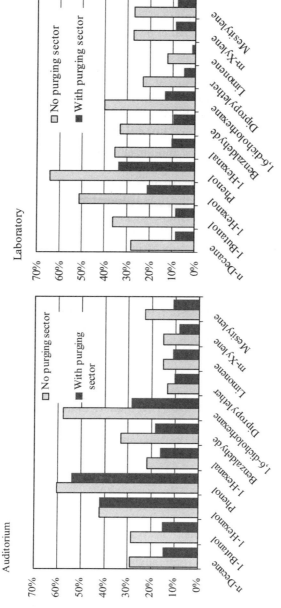

Fig. 10 Average volatile organic compound recirculation rates measured in both units, with and without a purging sector

- After 3 days: a preliminary chemical (TVOC) test and sensory irritation test with a panel of persons. If the TVOC concentration is higher than 5 mg/m^3 and/or more than 10% of a panel of persons perceive sensory irritation, no label is given.
- After 28 days: a final test comprising a toxicological test (LUR for all detected VOCs less than 10^{-5} and relevant compounds with concentrations more than 5 µg/m^3 should be evaluated toxicologically), TVOC test (less that 200 µg/m^3) and a sensory irritation test (odour/perceived air quality evaluation should have been performed). A label is given if all tests are passed.

Another labelling scheme is the indoor climate labelling scheme, in which building products are tested for their emission of VOCs and by a sensory evaluation of the emissions as a safety measure [62]. The parameter used for evaluation and as a criterion is the time required for the emission of VOCs of concern to decay to the point where their (modelled) room concentrations are below their indoor relevant values. These are based on 50% of either odour-threshold values or airway-irritation estimates.

The TVOC has been used as an indicator for IAQ as well as a label parameter for building products. It has even been used to evaluate the IAQ. However, the latest research indicates that the TVOC is an indicator for the presence of VOC indoors, but it can be used in relation to exposure characterisation and source identifications for VOCs only [63]. The TVOC cannot be used as an indicator for the presence of other pollutants and it cannot be used for normal regulatory risk assessment.

Besides the TVOC, the two units olf and decipol were introduced to quantify sensory source emissions (and label building products) and perceived air quality [15]. As mentioned before, the main item that is discussed with this method is the assumption that all pollutants have the same relation between exposure and response, i.e. that the calculated olf values from separate sources can simply be added.

References

1. Fanger PO (1972) Thermal comfort, analysis and applications in environmental engineering. McGraw-Hill, New York
2. Seifert B (2000) Ways to specify, reach and check guideline values for indoor air quality. Healthy buildings 2000, Helsinki, Finland, vol 4, pp 3–11
3. Berglund B, Bluyssen PM, Clausen G, Garriga-Trillo A, Gunnarsen L, Knöppel H, Lindvall T, Macleod P, Mølhave L, Winneke G (1999) Sensory evaluation of indoor air quality. Report no 20. European Collaborative Action Indoor air quality and its impact on man, EUR18676EN, Italy
4. Verein Deutscher Ingenieure (1986) VDI-Richtlinien, VDI 3881, Blatt 1/part 1. Olfactometry, odour threshold determination, fundamentals. Düsseldorf
5. Geldard A (1972) In: Geldard A (ed) The human senses, 2nd edn. Wiley, New York, chap 5
6. Kistiakowsky (1972) In: Geldard A (ed) The human senses, 2nd edn. Wiley, New York, chap
7. Amoore JE (1964) Sci Am February, pp 64–101

8. Dyson (1972) In: Geldard A (ed) The human senses, 2nd edn. Wiley, New York, chap
9. Dravniek (1968) In: Harper RE, Smith CD, Land DG (eds) Odour description and odour classification. J&A Churchill, London, pp
10. Mozell (1968) In: Harper RE, Smith CD, Land DG (eds) Odour description and odour classification. J&A Churchill, London, pp
11. Axel R (1995) Sci Am October, pp 130–137
12. Cain WS (1989) Perceptual characteristics of nasal irritation. NIVA course, Copenhagen, October 1989
13. Berglund B (1976) Psychol Rev 83:432–441
14. Cain WS, Cometto-Muniz JE (1993) Irritation and odour: symptoms of indoor air pollution. Proceedings of indoor air '93, vol 1, pp 21–31
15. Fanger PO (1988) Energy Buildings 12:1–6
16. Stevens SS (1957) Psychol Rev 64:153–181
17. Berglund B, Berglund U, Lindvall T (1976) Psychol Rev 83:432–441
18. Berglund B, Lindvall T (1990) Sensory criteria for healthy Buildings. Indoor air '90. Toronto, Canada, vol 5, pp 65–78
19. Bluyssen PM, Cornelissen HJM (1999) In: Design, construction and operation of healthy buildings, ASHRAE, pp 161–168
20. Garriga-Trillo A (1985) Funcion psicofisica y medida de la sensibilidad olfativa. Thesis. Universidad Autonoma de Madrid
21. Engen T (1972) In: Kling JW, Riggs Woodworth LA (eds) Schlosberg's experimental psychology, vol 1: Sensation and perception. Holt, Rinehart and Winston, New York, pp 11–46
22. Berglund B, Lindvall T (1979) Olfactory evaluation of indoor air quality. Indoor climate '78. Danish Building Research Institute, Copenhagen, pp 141–157
23. CEN (1994) Dynamic olfactometry to determine the odour threshold. Draft European preliminary standard, CEN TC264/WG2
24. Bluyssen PM, Walpot J (1993) Sensory evaluation of perceived air quality: a comparison of the threshold and the decipol method. Indoor air '93, Finland, vol 1, pp 65–70
25. Devos M, Patte F, Ronault J, Laffort P, van Gemert LJ (1990) Standardised human olfactory thresholds. IRL, New York
26. American Society for Testing and Materials (1981) Standard practices for referencing supra threshold odour intensity. Annual book for ASTM standards E544-75 (reapproved 1981), pp 32–44
27. Bluyssen PM, de Oliveira Fernandes E, Groes L, Clausen GH, Fanger PO, Valbjørn O, Bernhard CA, Roulet CA (1996) European project to optimize indoor air quality and erny corumption in office buildings. Indoor Air J
28. de Oliveira Fernandes E, Clausen G (1997) European database on indoor air pollution sources. Final report. Porto, Portugal
29. de Oliveira Fernandes E (2001) MATHIS publishable final report. Joule III programme. EC, Porto, Portugal
30. Bluyssen PM, Seppänen O, de Oliveira Fernandes E, Clausen G, Müller B, Molina JL, Roulet CA (2003) Why, when and how do HVAC systems pollute the indoor environment and what to do about it? Building and Environment, vol 38, issue 2
31. ASHRAE (1996) Standard 62-1989R proposed: Ventilation for acceptable indoor air quality. Appendix C: Air quality guidelines – informative. ASHRAE, Atlanta, USA
32. Gunnarsen L, Fanger PO (1992) Environ Int 18:43–54
33. Gunnarsen L, Bluyssen PM (1994) Sensory measurements using trained and untrained panels. Healthy buildings'94, Budapest, Hungary, vol 2, pp 533–538
34. Bluyssen PM (1990), Air quality evaluated by a trained panel. PhD study. Laboratory of Heating and Air Conditioning, Technical University of Denmark

35. Bluyssen PM (1991) Air Infiltration Rev 12:5–9
36. Bluyssen PM (1998) Protocol for sensory evaluation of perceived air pollution with trained panels. Internal project document 1.10, Delft, The Netherlands
37. Bluyssen PM, Elkhuizen PA (1994) Sensory evaluation of air quality: training and performance, part II. TNO report 94-BBI-R1664 (confidential)
38. Elkhuizen PA, Bluyssen PM, Groes L (1995) A new approach to determine the performance of a trained sensory panel. Healthy buildings '95, Milan, vol 3, pp 1365–1370
39. Bluyssen PM (1996) Methods and sensors to detect indoor air pollutants perceived by the nose. TNO report 96-BBI-R0873
40. Gardner JW, Bartlett PN (eds) (1992) NATO Advanced Study Institute Series 212
41. Elma K et al. (1989) Sens Actuators 18:291–296
42. Nakamoto T, Fukunishi K, Moriizumi T (1990) Sens Actuators B 1:473–476
43. Bruckman HWL et al. (1994) Kunsstof CO-sensor. TNO-industrie. TNO report 0795/U94
44. Gardner JW et al. (1990) In: Schild D (ed) Chemosensory information processing vol H39. Springer, Berlin Heidelberg New York, pp 131–173
45. World Health Organisation (1989) Indoor air quality: organic pollutants. EURO reports and studies no 11. WHO regional office for Europe, Copenhagen
46. European Collaborative Action (1994) Sampling strategy for volatile organic compounds (VOC) in indoor air. European Collaborative Action Indoor air quality and its impact on man, EUR 16051 en
47. van der Wal JW (1990) Stof in het binnenmilieu, Seminar, Maastricht, November 1990
48. Beck EM (1990) Filter facts. Indoor air '90, Toronto, Canada, July–August, vol 3, pp 171–176
49. Schmidt-Etkin D (1994) Biocontaminants in indoor environments. Indoor air quality update. Cutter Information Corporaton, USA
50. Lagoudi A, Loizidou M, Bernhars CA, Knutti R (1995) Identification of pollution sources that emit VOCs. Proceedings of healthy buildings '95, Milan, vol 3, pp 1341–1346
51. de Oliveira Fernandes E, Bluyssen PM, Molina JL (2001) SOPHIE, a European database on indoor air pollution sources, paper 1039. Air & Waste Management Association's Annual Conference & Exhibition, 24–28 June, Orlando
52. Cochet C, Kirchner S, De Bortoli M (1998) VOCEM – further development and validation of a small test chamber method for measuring VOC emissions from building materials and products. Final report
53. CEN ENV 13419 Building products – determination of emission of volatile organic compounds: part 1 – emission test chamber method; part 2 – emission test cell method; part 3 – procedure for sampling, storage of samples and preparation of test specimens
54. ISO DIS 16000-3: determination of formaldehyde; ISO DIS 16000-6: indoor air and emission test chamber air – determination of VOCs; active sampling on Tenax TA, thermal desorption and gas chromatography MSD/FID
55. Bluyssen PM (2004) A clean and energy-efficient heatury, ventilating and air-conditioning systems, Recommendations and advice, ISBN 90-5986-009-8, TNO Building and construction Research, Delft, The Netherlands
56. Bluyssen PM, Cox, Souto J, Müller B, Clausen G, Björkroth M (2000) Pollution from filters: what is the reason, how to measure and to prevent it? Healthy buildings 2000, Helsinki, August, vol 2, pp 251–256
57. Björkroth M, Müller B, Küchen V, Bluyssen PM (2000) Pollution from ducts: what is the reason, how to measure and how to prevent it? Healthy buildings 2000, Helsinki, August, vol 2, pp 163–168
58. Müller B, Fitzner K, Bluyssen PM (2000) Pollution from humidifiers: what is the reason, how to measure and to prevent it? Healthy buildings 2000, Helsinki, August, vol 2, pp 275–280

59. Roulet CA, Pibiri M-C, Knutti R (2000) Measurements of VOC transfer in rotating heat exchangers. Healthy buildings 2000, Helsinki, August, vol 2, pp 221–226
60. Roulet C-A, Pibiri M-C, Knutti R (2001) Effect of chemical composition on VOC transfer through rotating heat exchangers. CLIMA 2000, Naples, September
61. ECA (1997) Evaluation of VOC emissions from building products, solid flooring materials, report no 18. European Collaborative Action, Indoor air quality and its impact on man, environment and quality of life. EUR 17334 En, Italy
62. Wolkoff P, Nielsen PA (1996) Atmos Environ 30:2679–2689
63. Mølhave L (2003) Indoor Air 13(Suppl 6):12–19

The Handbook of Environmental Chemistry Vol. 4, Part F (2004): 219–239
DOI 10.1007/b94836
© Springer-Verlag Berlin Heidelberg 2004

Biomass Smoke and Health Risks – The Situation in Developing Countries

Kalpana Balakrishnan (✉) · Padmavathi Ramaswamy · Sambandam Sankar

Department of Environmental Health Engineering, Sri Ramachandra Medical College &
Research Institute, Porur, Chennai 600116, India
kalpanasrmc@vsnl.com

Abstract   About half of the world's population relies on traditional fuels such as biomass
(wood, agricultural residues, animal dung and charcoal) as the primary source of domestic
energy. Nearly 2×10^9 kg of biomass is burnt every day in developing countries. Use of open
fires for cooking and heating exposes an estimated 2 billion people to enhanced concentrations of particulate matter and gases, up to 10–20 times higher than ambient concentrations.
Recent studies estimate that exposure to indoor air pollution associated with household
solid fuel use may be responsible for nearly 1.6 million excess deaths in developing countries
and about 2.6% of the global burden of disease. An understanding of the linkages between
household fuel use and human health is especially crucial for developing strategies to
improve household environments and the status of public health as they form an important
prerequisite for all subsequent economic development. This chapter is devoted to describing the sources, emissions and patterns of exposure and consequent health risks for biomass
smoke associated with household fuel use in developing country settings. Potential research
needs in exposure and health risk assessments for addressing indoor air pollution and
household energy issues within the mainstream of environmental health and public health
policies of the region are also described.

Keywords   Indoor air pollution · Biomass fuels · Wood smoke · Developing countries ·
Rural environments

Abbreviations

ALRI   Acute lower respiratory infection
ARI   Acute respiratory infection
DALYs   Disability-adjusted life years
GM   Geometric mean
IGIDR   Indira Gandhi Institute for Development Research
$PM_{2.5}$ Particulate matter smaller than 2.5 μm
PM_{10} Particulate matter smaller than 10 μm
$PM_{3.5}$ Particulate matter smaller than 3.5 μm
USEPA   Unites States Environmental Protection Agency
WHO   World Health Organization

1
Background

About half of the world's population relies on traditional fuels such as biomass (wood, agricultural residues, animal dung and charcoal) as the primary source of domestic energy; nearly 2×10^9 kg of biomass is burned every day in developing countries [1, 2]. In developing countries such as India up to 86% of rural households and 24% of urban households currently rely on solid biomass fuels for their household energy needs [3] and the situation in other developing countries is similar (Fig. 1).

Household fuel demands have been shown to account for more than half of the total energy demand in most countries with per capita incomes under $1,000, while accounting for less than 2% in industrialized countries [4]. While it is known that as per capita incomes increase, households switch to cleaner, more efficient energy systems for their domestic energy needs (i.e. move up the "energy ladder"), these moves have largely been made owing to increases in affordability, the demand for greater convenience and energy efficiency.[1] With technological progress the income levels at which people make the transition to cleaner modern fuels has declined. However, in many rural areas despite the availability of cleaner fuels, they continue to use a combination of fuels as a result of socio-cultural preferences or as a risk reduction mechanism against an unreliable supply of cleaner fuels [6]. Household fuel generation, distribution and consumption are thus closely related to the overall structure of the energy, environmental and developmental systems that are operational in the respective countries.

---

[1] The energy ladder [5] is made up of several rungs with traditional fuels such as wood, dung and crop residues occupying the lowest rung. Charcoal, coal, kerosene, gas and electricity represent the next-higher steps sequentially. As one moves up the energy ladder, energy efficiency and costs increase, while typically the pollutant emissions decline. While several factors influence the choice of household energy, household income has been shown to be the one of the most important determinants. The use of traditional fuels and poverty thus remain closely interlinked.

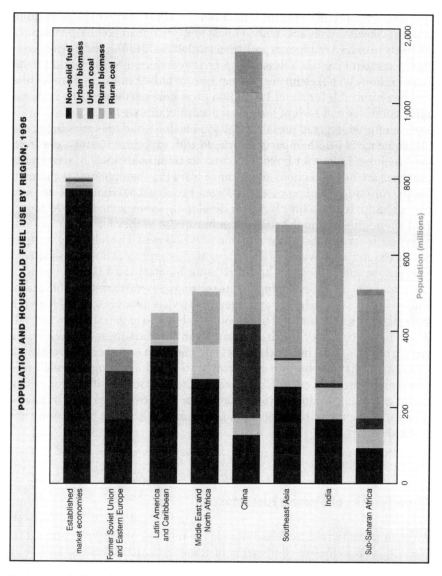

Fig. 1 Household fuel use across world regions (from World Assessment, United Nations Development Program [4])

The magnitude of environmental and health damage consequent to the widespread use of traditional solid fuels including biomass has only recently started to receive the attention of researchers and policymakers worldwide. Use of open fires for cooking and heating exposes an estimated 2 billion people to enhanced concentrations of particulate matter and gases, up to 10–20 times higher than ambient concentrations [7]. Although, biomass makes up only

10–15% of total human fuel use, since nearly half of the world's population cooks and heats their homes with biomass fuels, indoor air pollution exposures are likely to exceed outdoor exposures on a global scale [8].[2] The recently concluded comparative risk assessment exercise conducted by the World Health Organization (WHO) estimates that exposure to indoor smoke from solid fuels may be responsible for about 1.6 million premature deaths annually in developing countries and 2.6% of the global burden of disease [9].

Given the widespread prevalence of solid fuel use, and the emerging scientific evidence of health impacts associated with exposures to emissions from solid fuel use, indoor air pollution issues in rural households of developing countries are of tremendous significance from the standpoint of population health. An understanding of the linkages between household fuel use and human health is especially crucial for developing strategies to improve household environments and the status of public health as they form an important prerequisite for all subsequent economic development. The following sections of this chapter are devoted to describing the sources, emissions and patterns of exposure and consequent health risks for biomass smoke associated with household fuel use in developing country settings. Several examples from recent studies in these countries are described with a view not only to provide recent information on exposure and health impacts but also to describe special challenges in the conduct of such assessments in these settings. The concluding section offers insights into potential research needs in exposure and health risk assessments that may have an important bearing in addressing indoor air pollution and household energy issues within the mainstream of environmental health and public health policies of the region. The emissions and risks associated with the use of traditional household fuels other than biomass are not covered here.

2
Characteristics of Biomass Fuel Smoke

Air pollutants derived from biofuels are the result of incomplete combustion (conditions for efficient combustion of these fuels are difficult to achieve in typical household-scale stoves) and are practically the same for any type of biomass. However, the amount and the characteristics of pollutants produced during the burning of biomass fuels depend on several factors, including the composition of original fuel, combustion conditions (temperature and air flow),

---

[2] Exposures reflect concentrations that people are in contact with for specified durations. Since cooking takes places every day in homes at times when people are most likely to be home, the potential for individual exposure to indoor air pollutants is high compared with the situation of outdoor air pollution, where despite high concentrations exposures are low since people may not always be present where (and when) the pollution is high.

Table 1 Toxic pollutants from biomass combustion and potential for toxicity

Pollutant	Known toxicological characteristics
Particulates (PM_{10}, $PM_{2.5}$)	Bronchial irritation, inflammation, increased reactivity, reduced mucociliary clearance, reduced macrophage response
Carbon monoxide	Reduced oxygen delivery to tissues due to formation of carboxyhaemoglobin
Nitrogen dioxide	Bronchial reactivity, increased susceptibility to bacterial and viral lung infections
Sulfur dioxide	Bronchial reactivity (other toxic end points common to particulate fractions)
Organic air pollutants: Formaldehyde Acetaldehyde Phenols Pyrene Benzo[a]pyrene Benzopyrenes Dibenzopyrenes Dibenzocarbazoles Cresols	Carcinogenicity or co-carcinogenicity mucus coagulation, cilia toxicity, increased allergic sensitization, increased airway reactivity

Sources Smith [10], Cooper [11] and Smith and Liu [12].

mode of burning, and even the shape of the fireplace [10]. Hundreds of different chemical substances are emitted during the burning of biomass fuels in the form of gases, aerosols (suspended liquids and solids) and suspended droplets. These pollutants include carbon monoxide, small amounts of nitrogen dioxide, aerosols (called particulates in the air pollution literature) in the respirable range (0.1–10 μm in aerodynamic diameter), other organic matter, including polycyclic aromatic hydrocarbons such as benzo[a]pyrene, and other volatile organic compounds, such as benzene and formaldehyde. Smoke from wood-burning stoves has been shown to contain 17 pollutants designated as priority pollutants by the United States Environmental Protection Agency (USEPA) because of their toxicity in animal studies, up to 14 carcinogenic compounds, six cilia-toxic and mucous coagulating agents and four co-carcinogenic or cancer-promoting agents [11–13] (Table 1).

3
Indoor Air Pollutant Levels in Biomass Fuel Using Households –
Concentrations and Exposures

The majority of households in developing countries burn biomass fuels in poorly functioning earth or metal stoves or use open pits, often in an open fire configuration. Incomplete combustion in poorly ventilated kitchens thus results in very high levels of indoor air pollutants.[3] Some of the earliest studies to determine levels of indoor air pollutants associated with biomass combustion were carried out nearly 2 decades ago [14, 15]. Initial studies determined levels of total suspended particulates and exposures for cooks during cooking periods.[4] Subsequently many studies have been carried out for the determination of concentrations of other particulate fractions as well as other pollutants including CO, sulfur dioxide and nitrogen dioxide. Concentrations of total suspended particulates in the range 200–30,000 $\mu g/m^3$ and carbon monoxide concentrations between 10 and 500 ppm have been reported during the cooking period in some of the earlier studies [16–18]. Determinations of respirable particulate concentrations over 12–24 h have also been carried out that report 24-h means in the range 300–3000 $\mu g/m^3$ [19–21].

More recently systematic, large-scale 24-h measurements of respirable particulates have been conducted in Kenya, Guatemala and India, which in addition to pollutant concentrations have identified multiple household level determinants of concentrations and exposures. The Kenyan study [22] monitored 55 households for PM_{10} levels using continuous-monitoring light-scattering devices for 210 continuous days, addressed spatial and temporal variations in concentrations (that ranged from 200 $\mu g/m^3$ during noncooking periods far from the stove to 50,000 $\mu g/m^3$ during cooking close to the stove) and collected detailed time–activity records from 345 individuals to reconstruct individual daily average exposures. Adult women were exposed to the highest concentra-

---

[3] In many rural households of developing countries, it is common to find kitchens with limited ventilation being used for cooking and other household activities. Even when separated from the adjacent living areas, most offer considerable potential for the smoke to diffuse across. Use of biomass for space heating creates additional potential for smoke exposure in living areas.

[4] Exposures refer to the concentration of pollutants in the breathing zone during specific periods of time. Exposures reflect what is likely to be the internal body dose, the key determinant of health effects. Individual exposures would therefore be determined not only by the concentrations but also by how long the cooks spend breathing the polluted air. Women who cook, women who stay in close proximity to the stoves during cooking windows and young children who spend a considerable fraction of time with their mothers are thus likely to receive the highest exposures, while men despite living in the same households with high concentrations have lower exposures as they are less likely to be where the pollution is. Exposures are usually determined by attaching personal samplers to individuals or by determining concentrations in various household microenvironments together with detailed time budget assessments in these environments.

tions (24-h average exposures around 4,898 $\mu g/m^3$). Exposure levels for women were nearly 5 times higher than for men.

The Guatemalan study [23] determined 24-h $PM_{3.5}$ concentrations over an 8-month period for traditional and improved cooking stoves in 30 households. The 24-h concentrations ranged from 280 $\mu g/m^3$ to about 1,560 $\mu g/m^3$, with the improved stoves showing up to 85% reduction in concentrations.

In India, one of the earliest exposure assessment study was conducted in the households of Garhwal, Himalayas, and involved nearly 100 households in three villages across three seasons [19]. Twenty-four-hour exposures to total suspended particulates in the range of 250–1,690 $\mu g/m^3$ were reported with levels in winter up to 3 times higher than those in summer. More recently two large-scale exposure assessment exercises for respirable particulates have been completed in India in the southern states of Tamil Nadu and Andhra Pradesh. 436 rural households across four districts of Tamil Nadu were monitored for respirable particulates (median aerodynamic diameter of 4 μm) [24, 25]. Concentrations were determined during several cooking and noncooking windows in select clusters of households and were extrapolated to cover the entire region. Twenty-four-hour exposures were also calculated on the basis of these concentrations in conjunction with time–activity records of household members. Concentrations of respirable particulate matter ranged from 500 to 2,000 $\mu g/m^3$ during cooking in biomass-using households and average 24-h exposures ranged from 90±21 $\mu g/m^3$ for those not involved in cooking to 231±109 $\mu g/m^3$ for those who cooked. Twenty-four-hour exposures were around 82±39 $\mu g/m^3$ in households using clean fuels (with similar exposures across household subgroups).

Another recently completed study under the Energy Sector Management Assistance Programme of the World Bank [26, 27] in Andhra Pradesh quantified daily average concentrations of respirable particulates in 420 rural homes from three districts and recorded time–activity data from 1,400 household members. Mean 24-h average concentrations ranged from 70 to 850 $\mu g/m^3$ (geometric mean, GM, 56–570 $\mu g/m^3$) in gas-using versus solid-fuel-using households, respectively. Concentrations were significantly correlated with fuel/kitchen type and fuel quantity. Mean 24-h average exposures ranged from 75 to 443 $\mu g/m^3$. Amongst solid-fuel users mean 24-h average exposures were the highest for women cooks (GM 317 $\mu g/m^3$) and were significantly different from men (GM 170 $\mu g/m^3$) and children (GM 184 $\mu g/m^3$). Among women, exposures were the highest for women between the ages of 15 and 40 (most likely to be involved in cooking or helping in cooking), while among men, exposures were highest for men between the ages of 65 and 80 (most likely to be indoors). Fuel type, type and location of the kitchen and the time spent near the kitchen while cooking were the most important determinants of exposure across these households in southern India among other parameters examined, including stove type, cooking duration and smoke from neighbourhood cooking. The data is being used to calculate population exposures and develop a model to predict quantitative categories of exposure based on housing and fuel characteristics.

Some results from the studies just described are shown in Figs 2, 3 and 4 together with illustrations of the exposure situation in these households. A list of some recent studies carried out in developing countries is shown in Table 2, which compares the pollutants monitored, the spread in fuel/stove types, averaging periods for the monitoring and the concentrations in the households. The findings of these studies clearly show that rural women, children and men in biomass-using settings experience extremely high levels of particulates, gases and other noxious pollutants often an order of magnitude higher than what is considered safe levels of exposure in outdoor settings, for example.

The USEPA standards for 24-h PM_{10} and $PM_{2.5}$ concentrations in ambient environments are 150 and 65 $\mu g/m^3$, respectively. Indeed in some settings the levels exceed what is considered acceptable for even occupational exposures.[5] The threshold limit value adopted by the American Conference of Governmental Industrial Hygienists for respirable dusts is 5,000 $\mu g/m^3$ and for carbon monoxide is 29 mg/m^3. The comparison with these health-based standards (although currently available for settings other than indoor household environments) indicates the potential for significant health risks.

These preceding studies describe results from rural household settings; however, biomass use is not uncommon among the urban poor. These populations living in meager dwellings often on roadsides face dual risks from indoor and outdoor emissions. Community school programmes in India that provide free noon meals to children continue to use biomass fuels for cooking in kitchens often situated adjacent to classrooms. Exposures from biomass fuels are therefore not limited to rural household settings. Very limited information is currently available, however, to assess the scale and levels of such exposures.

Despite the growing number of studies, the database on indoor air pollutant concentrations and exposures is rather small compared with outdoor air pollution databases even within developing countries and certainly is much smaller than what is available in developed country settings. In the absence of nationally or internationally accepted standards for indoor air quality, the existing database of indoor air quality information, especially in rural household settings of developing countries, relies on findings of several independent research studies. Many of them have been carried out under considerable financial and logistic constraints that limit the number of pollutants/households that can be monitored and that necessitate the use of technologies that are largely dictated by local feasibilities. The absence of standards also results in large differences in methodologies and quality control mechanisms, making it difficult to extrapolate across studies.

---

[5] Occupational exposure limits are usually prescribed as a time-weighted average concentration for a normal 8-h workday and a 40-h workweek to which nearly all workers may be repeatedly exposed, day after day without adverse health effects. This limit is set for a healthy working-age population exposed only during the workweek and is clearly not applicable to the household setting with lifetime exposures for multiple subgroups including the most vulnerable, such as the aged and young children.

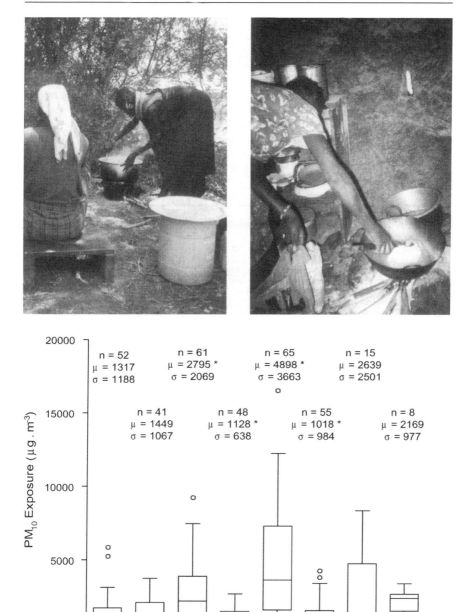

Fig. 2 Exposures to respirable particulates in rural households of Kenya. Source Ezzati et al. [22]. * The difference between male and female values is significant with P<0.0001

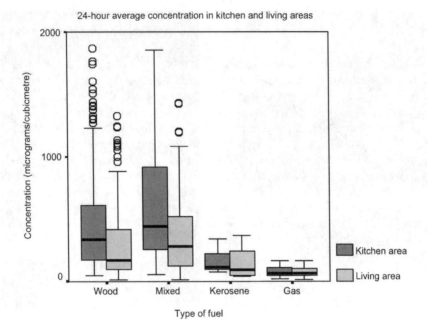

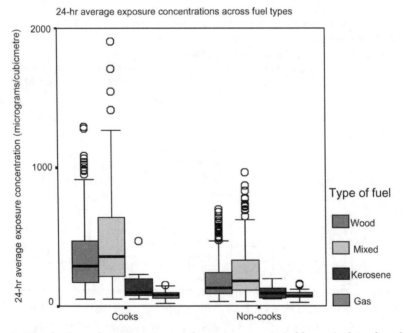

Fig. 3 Distributions of concentrations and exposures to respirable particulates from biomass smoke across rural households of Andhra Pradesh, India (mixed-fuel users are largely dung users). Source Balakrishnan et al. [26]

Fig. 4 The exposure situation in biomass-using households of southern India. Young girls are often involved in cooking and it is not uncommon for households to use the space for cooking and other household activities, creating additional exposure potentials for other members of the household. Source K. Balakrishnan

Table 2 Comparison of particulate levels as determined in some recent studies in developing countries. Total suspended particulates (*TSPs*), particulate matter (*PM*), respirable suspended particulate matter (*RSPM*), arithmetic mean (*AM*), geometric mean (*GM*)

Location and references	Fuel type	Average sampling duration	Time of measurement	Types of measurement (area or personal)	Areas of measurement (kitchen, living or outdoors)	Size fractions (TSP, PM_{10} or $PM_{2.5}$)	Exposure for adult women ($\mu g/m^3$)	Concentrations ($\mu g/m^3$)	
								AM	GM
Nepal (Davidson et al. 1986) [28]	Wood	1–2 h	Cooking period	Area	Kitchen	TSP RSPM<4 μm			8,800 4,700
Garhwal, India (Saksena et al. 1992) [19]	Wood/ shrubs		Cooking period			TSP			4,500
Pune, India (Smith et al. 1994) [20]	Wood	12–24 h		Area Personal		PM_{10} PM_{10}	1,100	2,000	
Mozambique (Ellegard 1996) [16]	Wood	1.5 h	Cooking period	Personal		PM_{10}	1,200		
Bolivia (Albalak et al. 1999) [29]	Dung	6 h	Total duration that the fire was on, including one cooking period	Area	Indoor kitchens Outdoor kitchens	PM_{10} PM_{10}		3,690 430	1,830 430
Kenya (Ezzati et al. 2000) [22]	Wood	24 h	Cooking period and time-activity recall	Area	Kitchen	PM_{10}	4,898		

Table 2 (continued)

Location and references	Fuel type	Average sampling duration	Time of measurement	Types of measurement (area or personal)	Areas of measurement (kitchen, living or outdoors)	Size fractions (TSP, PM$_{10}$ or PM$_{2.5}$)	Exposure for adult women (µg/m³)	Concentrations (µg/m³) AM	Concentrations (µg/m³) GM
Tamil Nadu, India (Balakrishnan et al. 2002) [24]	Wood	1–2 h	Cooking period	Personal	Kitchen	RSPM<4 µm	1,307		
	Agricultural waste						1,535		
	Wood	Daily average	Cooking period and time-activity recall	Personal and area	Kitchen, living and outdoor		226		
	Agricultural waste						262		
	Wood	1–2 h	Cooking period	Area	Living			847	498
	Agricultural waste							1,327	913
Guatemala (Albalak et al. 2001) [23]	Wood	24 h	24 h	Area	Kitchen	RSPM 3.5		1,930	1,560

Finally, given the complexity and the variability in the exposure setting with multiple household level determinants, household microenvironments, socio-cultural and behavioural determinants, there is a tremendous need to generate such environmental data on a regional basis.

4
Health Effects Associated with Biomass Fuel Smoke

Evidence for health effects associated with exposure to smoke from combustion of biomass fuels was provided initially by studies on outdoor air pollution as well as by studies dealing with exposure to environmental tobacco smoke. Criteria documents for outdoor air pollutants published by the USEPA [13], for example, detail the effects of many components, including particulate matter, carbon monoxide, oxides of sulfur and nitrogen and polycyclic aromatic hydrocarbons.

Considerable scientific understanding now exists about the aerodynamic properties of the particles that govern their penetration and deposition in the respiratory tract. The health effects of particles deposited in the airway depend on the defence mechanisms of the lung, such as aerodynamic filtration, mucociliary clearance and in situ detoxification. Since most particulate matter in biomass fuel smoke is less than 3 μm in diameter, it is possible that such particulate matter may reach the deepest portions of the respiratory tract and alter defence mechanisms. Several biomass fuel combustion products may also impair the mucociliary activity and reduce the clearance capacity of the lung, resulting in an increased residence time of inhaled particles, including microorganisms, and favours their growth. In situ detoxification, the main mechanism of defence in the deepest nonciliated portions of the lung, may also be compromised by exposure to components of biomass fuel smoke [30].

Gases such as carbon monoxide are known to bind to haemoglobin thus reducing oxygen delivery to key organs and may have important implications for pregnant women, with developing fetuses being particularly vulnerable. Although emissions of other gases such as sulfur dioxide and nitrogen dioxide are of lesser concern in biomass combustion (very high levels of sulfur dioxide may be reached with other solid fuels such as coal), they are known to increase bronchial reactivity. Polycyclic aromatic hydrocarbons such as benzo[a]pyrene are known carcinogens.

On an epidemiological platform, the earliest evidence linking biomass combustion, indoor air pollution and respiratory health came from studies carried out in Nepal and India in the mid 1980s [31, 32]. Since then there has been a steady stream of studies linking biomass combustion and several health effects especially in women who cook with these fuels and young children. Recent reviews describe the evidence linking exposures from biomass combustion and health outcomes, including acute lower respiratory infections (ALRIs), chronic obstructive lung disease, tuberculosis, perinatal outcomes, including low birth weight, eye disease and cancer [33–35]. Important observations made in these reviews are presented here.

ALRIs are the single most important cause of mortality in children aged under 5 years and account for nearly 2 million deaths annually in this age group [36]. Associations between biomass fuel exposure and ALRIs were documented as early as 1968 [37]. However the criteria for diagnosis have been revised over the years and many of the earlier studies were unable to clearly define outcomes. Only about a dozen studies are included in the recent reviews that have been conducted in recent years in developing country settings (and one in developed country settings amongst Navajo reservations in the USA) confirming the rigorous case-selection criteria of the WHO [38–40]. Most studies showed increased odd ratios (between 2 and 3) although a few showed no association.[6] While most of these studies have fairly well defined health outcomes, they have relied on proxy exposure indicators such as the use of biomass fuels, the time spent near the stove or if child is carried on the back. As illustrated in the preceding section, while the use of biomass fuels results in substantially higher concentrations in these households compared with those using gas or other modern fuels, intrahousehold differences can be substantial. Differences in individual exposures can be masked significantly by such binary classification schemes and result in ambiguity in health outcome assessments. Despite these uncertainties, the degree of consistency in terms of increased risks across exposure settings indicates a substantial burden of disease-attributable ALRI owing to biomass smoke exposure (see Sect. 5).

A recent study examined the exposure–response relationship between biomass combustion and ARI in children of rural Kenyan households [41]. This was preceded by rigorous quantitative exposure assessments in the same households [22]. Quantitative exposure assessments are therefore crucial for the development of exposure–response relationships and greatly facilitate subsequent health risk assessments.

Many studies have reported an association between exposure to biomass smoke and chronic obstructive pulmonary disease [42–44]. Cigarette smoking accounts for 80% of the incidence of chronic bronchitis in non-biomass-using settings. The incidence of chronic bronchitis in women (who are largely non-smokers) has been reported to be comparable to or sometimes higher than that in men (including smokers) in biomass-using households, indicating positive associations with smoke exposure. Unlike ARIs, obstructive pulmonary disease is a chronic outcome and is less subject to confounding and studies have

---

[6] Odds ratios represent the ratio of the probability of occurrence of an event to nonoccurrence; for example, an elevated odds ratio in biomass-using households reflects the incremental risks for people in this set of households compared with risks for people in clean-fuel-using households. An odds ratio of 2 for acute respiratory infection (ARI) in children for biomass-using households, for example, would imply a twofold higher risk of ARI for these children compared with the reference group of children in clean fuel (gas) using households.

confirmed very high particulate levels in these households. Risk factors for chronic obstructive pulmonary disease associated with tobacco smoking include bronchial hyper-reactivity, atopy and genetic susceptibility, all of which could apply to biomass exposure [45]. These considerations together indicate the substantial burden of obstructive pulmonary diseases (especially for women) in biomass-using households.

Biomass exposure has also been shown to be associated with other health outcomes, including blindness (cataracts), self-reported tuberculosis, reduced birth weight and increased perinatal mortality but the evidence for such associations is not as strong as what is available for ARI and chronic obstructive pulmonary disease. They are subject to even greater uncertainties in both the exposure and the health assessment components and the reader is referred to the references cited earlier for individual studies concerned with particular health risks.

5
Global Burden of Disease Attributable to Indoor Air Pollution

The global burden of disease methodology [46] allows one to calculate the health burden in terms in terms of a common metric, disability-adjusted life years (DALYs) – the number of healthy life expectancy years lost because of a disease or a risk factor. It has been applied to indoor air pollution owing to the use of solid fuels both globally and nationally in India [7, 47, 48]. Although on a global platform this includes risks from traditional fuels other than biomass (most importantly coal), risks from biomass are a significant portion of it, given the much higher prevalence of use in most less developed countries (Fig. 1).

The results of such calculations are summarized in Fig. 5 and they are compared with the burden from outdoor air pollution across world regions. These assessments using rather conservative approaches, relying largely on epidemiological evidence from studies carried out in developing countries, estimate a very high global burden attributable to indoor air pollution, up to 2 million excess deaths and 4% of all DALYs. These risks are comparable to risks from tobacco and are only exceeded by malnutrition (16%), unsafe water and sanitation (9%) and unsafe sex (4%). In developing countries (which bear the largest share of this burden) this creates additional challenges, increasing pressure on scarce resources for mitigation of a growing number of risk factors. A more recent WHO analysis for year 2000 done as part of the global comparative risk assessment exercise has determined slightly smaller risks, but they lie in the same range (9).

As the figure also suggests, the disease burden consistently falls as the region develops and incomes grow, reflecting the need for addressing indoor air pollution in the mainstream of poverty alleviation initiatives. ARI is the leading cause of disease burden in children under 5 years and it contributes the most (up to 80%) to deaths and DALYs attributable to indoor air pollution in most

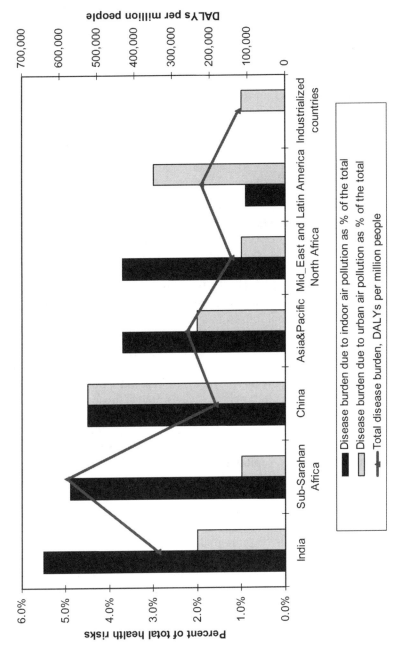

Fig. 5 The burden of disease attributable to indoor air pollution across world regions. Source World Bank [48]

biomass-using settings, thereby also indicating the need to address this issue within mainstream children's health initiatives. Finally, women who are at the centre of care giving at the family level bear a significant disease burden that can have implications beyond their own health (most importantly, children's health). Health risks from indoor air pollution in household settings thus have complex interlinkages and an understanding of these linkages in a holistic manner is crucial for the design of strategies to minimize the same.

6
Options for Interventions

Household-level energy interventions have been in the past largely centred on providing improved fuel efficiency either by using better fuels (which necessarily use stoves with higher combustion and heat transfer efficiency) or by using improved stoves with the same fuels [1]. Since fuel substitution has not been a feasible option for most communities in less developed countries owing to the high costs or nonavailability, improved stoves have remained the most common intervention. Emissions and health risks from associated exposures have, however, seldom been a part of technological considerations in the design of interventions thus far.[7] In India and China, where national-level improved stove programs have been operational for the last 2 decades, the impact of the programs has largely been assessed in terms of the number of units disseminated and the impact on health risk reduction remains poorly understood. The residual pollutant levels even amongst those in regular use are often high [21, 29, 49, 50].

Many considerations have been shown to determine the acceptability of improved stoves among communities, including market environments, people's socio-cultural preferences, capital and running costs and access to wood for fuel. People living in rural settings are often under limited pressure to conserve wood (except in arid areas) and are not willing to spend any additional resources on alternative fuels or stoves. Indeed when wood for fuel becomes scarce people move down the energy ladder, shifting to agricultural produce and animal dung, known to be more polluting. Perception of health risks thus currently plays no (or a very limited) role in determining energy choices that households make in their daily lives. Interventions that reduce exposures either directly (through behavioural interventions) or indirectly (promote dispersion through improved ventilation/housing design) have been described [51] but

---

[7] Improved efficiency is not always accompanied by reduced emissions [45]. An increase in efficiencies may be accomplished by increasing merely the heat transfer efficiency, in which case emissions (which are directly related to fuel combustion) are not reduced. Further, expected gains in efficiency (set in laboratory conditions) are seldom realized under field conditions, with most improved stoves resulting in savings in fuel consumption of less than 25%.

evidence for exposure and health risk reductions are not currently available from such measures. Strengthening this evidence would enable better and perhaps more objective comparisons to be made across interventions and to be advocated on the basis of relative health economic and environmental benefits.

7
Challenges and Opportunities for the Future

The burden of environmental health risks is just beginning to catch the attention of health policymakers in developing countries. Developing countries are faced with several new environmental challenges against a backdrop of traditional public health risks. Allocation of scant resources for risk reduction will necessarily demand that the weight of evidence for each of these risk factors be built on a strong scientific foundation. The preceding account of studies that contribute towards better estimating the health risks associated with biomass exposures points out the need for augmenting quantitative assessments of health risks, which in turn rest on obtaining better exposure estimates.

Exposures to indoor air pollutants associated with household fuel combustion happens every day in homes that differ widely in their configurations and, given the multiple household level determinants of individual exposure that vary across world regions and within countries, there is a need to collect information on both exposures and their determinants on a regional basis. The challenge subsequently would be to identify a key set of exposure determinants that would provide sufficient resolution, to classify populations into exposure subcategories. The currently available techniques for the conduct of such assessments are laborious, expensive and there is a need to develop newer methods that are suited for being scaled up to local, national and regional levels.

The exposure and the health studies on this issue have largely remained separate from each other. While financial constraints may be responsible for some studies not being able to address them simultaneously, it is also in some measure a reflection of the lack of capacities in performing quantitative environmental health assessments in developing country settings. Even in instances where health-based environmental standards are available (e.g. criteria outdoor air pollutants) they are based on underlying exposure–response relationships that are largely derived from developed country studies. Risk perception and risk communication mechanisms within research/policy communities are therefore significantly handicapped either owing to the lack of locally derived relationships, which reduces acceptability, or to the lack of understanding of methodologies, which limits transferability across settings. With indoor air pollution largely being a developing-country issue with strong regional differences, it is anticipated that health-based standards will have to rely on studies largely executed in individual countries. The strengthening of local technical capacities through academic and interagency partnerships is thus crucial to enhance not only the cost-effectiveness of research initiatives but also to ensure sustainability of subsequent environmental management initiatives and supporting policies.

The issue of indoor air pollution associated with household fuels in developing countries is deeply embedded in a matrix of environmental, energy, health, and economic/developmental considerations. An in-depth understanding of the potential for health risks is therefore crucial for ensuring that the most vulnerable poor communities amongst us are not required to endure years of suffering, before development can "catch up" with them. Indeed if human development is the goal, addressing health risks is an important mechanism of ensuring equity in quality of life amongst populations and it is hoped that the information presented here represents a small incremental step towards achieving the same.

Acknowledgements The studies on exposure assessment in southern India reported here were funded through a grant from the World Bank under the Energy Sector Management Assistance Programme and the United Nations Development Program Capacity 21 project through the Indira Gandhi Institute for Development Research (IGIDR). We thank Kirk Smith at the University of California, Berkeley, Jyothi Parikh, and Kirit Parikh at the IGIDR and Kseniya Lvovsky and Priti Kumar at the World Bank for their contributions in these studies as well as for providing valuable guidance in many of our activities relating to indoor air pollution in India. We also thank the Chancellor, the Vice Chancellor and the management of Sri Ramachandra Medical College and Research Institute for their support. Finally, our grateful thanks to the members of the households who graciously permitted us to use their homes for the studies reported here.

References

1. Barnes DF, Openshaw K, Smith KR, Van der Plas R (1994) World Bank technical paper. Energy series 242. World Bank, Washington, DC
2. Reddy AKN, Williams RH, Johansson TB (1996) Energy after Rio: prospects and challenges. United Nations, New York
3. Government of India (1991) Census of India. Office of the Registrar General and Census Commissioner, New Delhi
4. United Nations Development Programme (2000) World energy assessment: energy and the challenge of sustainability. UNDP, New York
5. Reddy AKN, Reddy BS (1994) Energy 19:561
6. Masera OR, Saatkamp DB, Kammen DM (2000) World Dev 28(12):2083–2103
7. World Health Organization (1999) Guidelines for air quality. WHO, Geneva
8. Smith KR(1988) Environment 30:16–34
9. World Health Organization (2002) World health report: reducing risks, promoting healthy life. WHO, Geneva
10. Smith KR (1987) Biomass fuels, air pollution and health. A global review. Plenum, New York
11. Cooper JA (1980) J Air Pollut Control Assoc 30:855
12. Smith KR, Liu Y (1993) In: Samet JM (ed) Epidemiology of lung cancer: lung biology in health and disease. Dekker, New York, pp
13. United States Environmental Protection Agency (1997) Revisions to the national ambient air quality standards for particulate matter. Federal Register 62:38651

14. Smith KR, Aggarwal AL, Dave RM (1983) Atmos Environ 17:2343
15. Aggarwal AL, Raiyani CV, Patel PD et al (1982) Atmos Environ 16:867
16. Ellegard A (1996) Environ Health Perspect 104:980
17. Pandey MR, Neupane RP, Gautam A, Shrestha IB (1990) Mountain Res Dev 10:313
18. Reid H, Smith KR, Sherchand B (1986) Mountain Res Dev 6:293
19. Saksena S, Prasad R, Pal RC, Joshi V (1992) Atmos Environ 26A:2125
20. Smith KR, Apte MG, Yoqing M, Wongsekiarttirat W, Kulkarni A (1994) Energy 19:587
21. McCracken JP, Smith KR (1998) Environ Int 2:739
22. Ezzati M, Saleh H, Kammen DM (2000) Environ Health Perspect 108:833
23. Albalak R, Bruce N, McCracken J, Smith KR, Gallardo T (2001) Environ Sci Technol 35:2650
24. Balakrishnan K, Parikh J, Sankar S, Padmavathi R, Srividya K, Vidhya V, Prasad S, Pandey V (2002) Environ Health Perspect 110:1069
25. Parikh J, Balakrishnan K, Pandey V, Biswas H (2001) Energy 26:949
26. Balakrishnan K, Sankar S, Padmavathi R, Arnold J, Mehta S, Smith KR, Rao C, Kumar S, Kumar P, Akbar S, Lvovsky K (2002) Proceedings of the 9th international conference on indoor air quality and climate, Monterey, USA, p 4:560
27. Balakrishnan K, Sankar S, Padmavathi R, Mehta S, Smith KR (2003) J Environ Sci Pol 5:87
28. Davidson CI, Lin S, Osborn JF(1986) Environ Sci Technol 20:561–567
29. Albalak R, Keeler G J, Frisancho AR, Haber M (1999) Environ Sci Technol 33:2505–2509
30. Demarest GB, Hudson LD, Altman LC (1979) Am Rev Respir Dis 119:279
31. Pandey MR (1984) Thorax 39:331–36
32. Ramakrishna J, Durgaprasad MB, Smith KR (1989) Environ Int 15:341
33. Bruce N, Perez-Padilla, Albalak A (2000) Bull WHO 78:1078
34. Smith KR (1993) Annu Rev Energy Environ 18:529
35. Smith KR, Samet JM, Romieu I, Bruce N (2000) Thorax 55:518
36. Murray CJL, Lopez (1996) The global burden of disease. WHO, Harvard University Press, MA
37. Sofuluwe GO (1968) Arch Environ Health 16:670
38. Armstrong JR, Campbell H (1991) Int J Epidemiol 20:424
39. Bruce N, Neufeld L, Boy E, West C (1998) Int J Epidemiol 27:454
40. Robin LF, Lees PSJ, Winget M, Steinhoff M, Moulton LH, Santhosham M, Correa A (1996) Pediatr Infect Dis J 15:859
41. Ezzati M, Kammen DM (2001) Environ Health Perspect 109:481
42. Albalak R, Frisancho AR, Keeler GJ (1999) Thorax 54:1004–1008
43. Norboo T (1991) Int J Epidemiol 20:749–757
44. Padmavati S, Pathak SN (1959) Circulation 20:343
45. Anderson HR (1979) Int J Epidemiol 8:127–135
46. Murray CJL, Lopez A (1997) Lancet 349:1436
47. Smith KR (2000) Proc Natl Acad Sci USA 97:13286
48. World Development Report (2000) The World Bank, Washington, DC
49. Ezzati M, Mbinda MB, Kammen DM (2000) Environ Sci Technol 34:578
50. Mumford J, He X, Chapman R (1987) Science 235:217
51. Ballard-Tremeer G, Mathee A (2000) USAID/WHO global technical consultation on the health impacts of indoor air pollution and household energy in developing countries, Washington, DC

The Handbook of Environmental Chemistry Vol. 4, Part F (2004): 241–263
DOI 10.1007/b94837
© Springer-Verlag Berlin Heidelberg 2004

Strategies for Healthy Indoor Environments – a Chinese View

Jiming M. Hao (✉) · Tianle L. Zhu

Department of Environmental Science and Engineering, Tsinghua University, China
hjm-den@tsinghua.edu.cn

Abstract   The characteristics of indoor air pollution are reviewed. Three sources of indoor air pollution are of great importance for the health of people indoors: fuel combustion, indoor decorating and refurbishing materials, and building materials in China. The principal pollutants from these sources are combustion products, cooking fumes, formaldehyde, volatile organic compounds, ammonia, and radon. The reasons for indoor air quality problems are analyzed. Strategies and measures of controlling indoor air pollution are summarized. They include constituting standards and guidelines of indoor air pollution control, framing management policies of controlling indoor air pollution from energy consumption and civil building engineering work as well as other measures such as the establishment of administration, special monitoring teams, research and development, education and training.

Keywords   Indoor air pollution · Characteristics · Strategy · China

1
Introduction

In China, the study of the relationship between indoor air quality and human health began in the 1980s. According to investigations and monitoring, the main indoor air pollutants at that time were fuel combustion products, cooking fumes, and environmental tobacco smoke. Epidemiological studies showed that poor indoor air quality caused physiological dysfunction or diseases. However, adequate attention was not given to the problem because of the poor awareness of the importance of indoor air quality, the relative infrequency and multifactorial nature of some chronic respiratory diseases, and inadequate monitoring data.

In recent years, indoor decoration and refurbishment of buildings have become a focus of consumer consumption as personnel incomes have risen, and ownership of homes has become possible as a result of China's housing reform (state-owned or enterprise-owned residential houses in urban areas are sold to residents). Statistics indicate that the production value growth in the indoor decoration and refurbishment industry over the past 5-year period has been dramatic, with average annual increases of 20%. The total product value from the sector reached ¥ 660 billion in 2001 [1]. On the other hand, the indoor air quality has steadily deteriorated as a result of misusing decoration and refurbishment materials containing harmful substances. Significant amounts of formaldehyde, volatile organic compounds (VOCs), ammonia and radon are being released into indoor environments, thus posing a hazard to human health. Reports of poisoning and even death from indoor decoration and refurbishment often appear in the news, and the sick building syndrome is becoming prevalent. Thus, the indoor air quality problem is causing deep concern to the government and the public.

In this chapter, characteristics of indoor air pollutants in China are introduced first. Then the main reasons for the indoor air quality problem are analyzed. Finally, current existing strategies and measures for reduction of indoor air pollution are described.

2
Characteristics of Indoor Air Pollution in China

There are many sources of indoor air pollution in any home. These include (1) occupants and their activities, (2) building, indoor decorating, and refurbishing materials, (3) appliances, office equipment, and supplies, and (4) outdoor air. The relative importance of any single source depends on how much of a given pollutant it emits and how hazardous those emissions are. At present, residential heating and cooking activities, indoor decorating, and refurbishing materials are the major sources of indoor air pollution in China.

2.1
Indoor Air Pollution from Residential Heating and Cooking

In China, energy used by households is an important component of China's energy consumption. Coal, biomass, and gas are major energy types for cooking and space heating (Table 1) [2]. Burning such fuels produces large amounts of air pollutants, including particulates, sulfur dioxide, nitrogen dioxide, carbon monoxide, and carbon dioxide, in the confined space of the homes. Exposure to pollutants is often far higher indoors than outdoors because of insufficient ventilation. Recently, a series of studies done in China has shown that residential energy-related indoor air pollution is very serious [3–8].

Compared with using gas, using coal produces inhalable particulates, sulfur dioxide, nitrogen oxides, and carbon monoxide with higher concentrations in kitchens (Table 2) [9]. The research results from Shanghai and three other cities also show that children who belong to coal-consuming families have more respiratory disease symptoms and a higher prevalence of respiratory disease than those who belong to gas-consuming families (Table 3) [9].

Table 1 Annual energy consumption for nonproduction purposes by variety [2]

Item	1990	1995	1998	1999
Total (10^4 tce)[a]	15,800	15,745	14,393	14,552
Coal (10^4 t)	16,700	13,530	8,884	8,408
Liquefied petroleum gas (10^4 t)	159	534	769	878
Natural gas (10^8 m^3)	19	19	24	26
Coal gas (10^8 m^3)	29	27	74	81
Heat (10^{10} kJ)	8,972	12,637	18,711	20,127
Electricity (10^8 kWh)	481	1,006	1,325	1,481
Kerosene (10^4 t)	105	64	63	71
Biomass	Stalks (10^4 tce)		15,092	12,280	12,502
	Firewood (10^4 tce)		10,013	8,401	7,791

[a] Not including biomass energy.

Table 2 Concentrations of pollutants in kitchens using coal and liquefied petroleum gas in urban China [9]

Fuel	Average value ± standard deviation (mg/m^3)				
	SO_2	NO_2	CO	Inhalable particulates	CO_2
Coal	0.53±0.34	0.09±0.08	26.47±21.8	0.64±0.55	0.76±0.57×10^4
Liquefied petroleum gas	0.25±0.19	0.08±0.07	0	0.26±0.20	0.88±0.50×10^4

Table 3 Relationship of fuel usage and incidence rate of respiratory diseases and their symptoms expressed as a percentage of the affected population [9]

Disease or symptom	Chende		Shenyang		Shanghai		Wuhan	
	Coal	Gas	Coal	Gas	Coal	Gas	Coal	Gas
Cough	57.6	49.2	27.6	25.5	72.0	28.2	26.2	26.8
Too much sputum	24.7	17.0	6.6	4.6	40.2	15.9	7.9	7.4
Tracheitis	7.2	5.8	2.0	1.1	21.9	2.3	4.2	4.4
Pharyngitis	6.9	4.8	3.2	2.7	15.4	3.3	7.5	7.0
Tonsillitis	15.4	16.0	12.4	10.9	16.7	7.0	10.3	13.1
Asthma	0.8	0.8	0.9	0.0	1.6	1.3	0.5	0.6
Congestion of throat	4.4	6.3	24.7	20.3	29.6	18.3	24.8	19.7
Retropharyngeal lymph	19.5	18.8	18.1	14.4	25.4	26.9	60.3	50.3
Folliculosis	8.2	806	6.3	3.2	10.9	2.3	27.6	20.7
Nose-excretion increase	8.2	8.6	6.3	3.2	10.9	2.3	27.6	20.7
Nose hyperonychosis	3.6	4.6	8.9	4.3	9.0	3.3	21.0	12.1
Total number of people	389	394	348	439	311	301	214	503

The indoor air pollution contributed by biomass combustion is similar to or even higher than that contributed by burning coal because of using simple stoves, of which the average thermal combustion efficiency is only about 10–15%. Comprehensive research in rural Xianwei county has shown that firewood smoke contains many kinds of carcinogenic air pollutants (Table 4) [10].

Using biomass as a fuel is often coupled with inefficient ventilation; thus, extremely high levels of indoor air pollutants are produced. The outcome is that people – mainly women and children in rural areas and urban slums who spend most time in their homes – are exposed to high levels of indoor air pollution. It is estimated that nearly 2 million women and children die every year in developing countries as a result. About half of these deaths occur in India and China [11].

Table 4 Air pollutants in firewood smoke [10]

Pollutant	Emission factor (g/kg)		
	Above oven (1)	Around oven (2)	Ratio above oven/ around oven
Fluorine	0.020	0.0047	4.3
Anthracene/phenanthrene	0.006	0.0088	10.9
Phenol	0.1	0.02	5.0
Fluorane	0.022	0.0016	13.7
Pyrene	0.019	0.0016	11.9
Benzanthracene	0.0177	0.0019	9.3
Benzo[j]fluoranthene	0.009	0.0015	8.0
Benzo[ghi]pyrene	0.0059	0.0014	4.2
Benzo[a]pyrene	0.0025	0.00073	3.4
Dibenzanthracene	0.0010	0.00018	5.6
Dibenzene	0.041	0.0091	4.5
Benz[c]phenanthrene	0.0025	0.008	0.31
Dibenzopyrene	0.0007	0.0004	1.8
Formaldehyde	0.2	0.4	0.5
Propionaldehyde	0.2		
Acetaldehyde	0.1		
Isobutyraldehyde	0.3	0.5	0.6
Methyl phenol	0.2	0.06	3.3
o-Dihydroxybenzene	0.01	0.014	0.7

Table 5 Comparison of concentration of indoor pollutants in kitchens and bedrooms [12]

	Sample number	PM_{10} ($\mu g/m^3$)	SO_2 ($\mu g/m^3$)	CO (mg/m^3)
Kitchen	373	518±27	12.4±36	2.0±9.9
Bedroom	504	340±9	10.9±18	1.62±6.0

Fuel combustion generally occurs in kitchens, so the concentrations of pollutants in kitchens are higher than those in bedrooms (Table 5) [12]. With regard to different kinds of heating systems, separate ones (with small coal stoves) in individual residence units have higher indoor pollution than central heating systems in winter (Table 6) [13].

Perhaps the most compelling example of the health impact from indoor air pollution contributed by coal users in households is the extremely high lung cancer rates among nonsmoking women in rural Xianwei country. The three communes of this county, in Yunnan, have the highest prevalence of lung cancer in China. The age-adjusted lung cancer mortality rate between 1973 and 1979 was 125.6 per 100,000 women, compared with average rates of 3.2 and 6.3 for Chinese and USA women, respectively, for the same time. Because surveys

Table 6 Effect of heating method on concentration of indoor pollutants [13]

	Inhalable particulates (mg/m^3)		SO$_2$ (mg/m^3)		NO$_2$ (mg/m^3)		CO (mg/m^3)	
	Summer	Winter	Summer	Winter	Summer	Winter	Summer	Winter
Separate (or individual) heating system								
Kitchen	0.071	0.682	0.092	0.554	0.037	0.070	2.95	8.24
Bedroom	0.057	0.286	0.079	0.397	0.032	0.053	2.31	7.96
Central heating system								
Kitchen	0.074	0.262	0.072	0.180	0.045	0.087	2.54	5.86
Bedroom	0.048	0.127	0.042	0.154	0.034	0.055	1.60	5.33

showed that virtually no women (in the county) smoked tobacco products, other sources of potent exposure must have contributed to these troubling rates. Analyses of indoor air and blood samples from the women indicate that fuel burning inside their homes was largely responsible for the lung cancers. The studies found a strong association between the existence of lung cancer in females and the duration of time spent cooking food indoors. The levels of carcinogenic compounds present in smoky coal (a local type of coal that smokes copiously) were found to be much higher in the women who used smoky coal for cooking [14].

In addition to combustion products, cooking fumes are also major pollutants that cause indoor air pollution. Stir-fry and deep-fry are the main Chinese-style cooking methods. When heated to high temperatures in woks, cooking oils may emit harmful fumes containing aldehydes, ketones, hydrocarbon compounds, fatty acids, alcohols, aromatic compounds, and heterocyclic compounds. These compounds potentially cause diseases of the lung, the liver, and the immune system. Researchers from China and Canada found that cooking in a nonseparate kitchen, heating oils to high temperatures, smokiness in the kitchen while cooking, use of rapeseed oil to stir-fry, and frequent stir-frying increased the risk of lung cancer in a group of nonsmoking women living in Shanghai. Frying in hot oil is associated with a 1.6-fold increase in lung cancer risk. The highest risks were associated with the use of rapeseed oil and high temperatures [15].

Gao et al. [16] investigated the cause of death of 2,345 workers who were employed in cooking. The result showed that the lung cancer mortality rate was 98.02 per 100,000 workers, significantly higher than the average lung cancer mortality rate of local residents. In addition, the epidemiological study also indicated that people working in kitchens preparing dishes, compared with those dealing with rice and flour, risk having a higher lung cancer mortality rate because of exposure to heavier fumes over longer periods.

2.2
Indoor Air Pollution from Indoor Decorating and Refurbishing Materials

In China, indoor air pollution is becoming more and more serious because of the use of poor-quality decorating and refurbishing materials and processes. The main harmful substances involved include formaldehyde and VOCs. They are regarded as "invisible killers" in China.

Recent statistics from Beijing's Chemical Poisonous Substances Test Centre indicate that more than 400 poisoning accidents reported annually are the direct result of indoor air pollution caused by construction and interior decorating materials. The statistics also revealed that some 10,000 people suffer from relevant accidents on an annual basis in Beijing [17].

The most significant sources of indoor formaldehyde are likely to be pressed wood products made using adhesives that contain urea–formaldehyde resins. Pressed-wood products made for indoor uses include particleboard (used as subflooring and shelving and in cabinetry and furniture), hardwood plywood paneling (used for decorative wall coverings and used in cabinets and furniture), and medium-density fiberboard (used for drawer fronts, cabinets, and furniture tops). The demand for pressed-wood products is steadily increasing as individual incomes rise, which can give a rough idea of the consumption of wood adhesives in China (Table 7) [18].

In China, formaldehyde emission of the pressed-wood products sold is generally high. A test showed that the average formaldehyde release rates of four kinds of typical wood-based panels, i.e., particle board, medium-density

Table 7 Estimated consumption of wood adhesives in China units: $\times 10^3$ t [18]

Type	1994		2000		2010	
	Urea–formaldehyde	Phenol–formaldehyde	Urea–formaldehyde	Phenol–formaldehyde	Urea–formaldehyde	Phenol–formaldehyde
Plywood						
Interior	126.3		165.0		215.0	
Specialty		12.6		20.0		27.6
Bamboo		4.4		9.0		12.0
Particle board	131.2		183.0		304.0	
Fiberboard						
Hardboard		15.7		13.5		14.5
Medium-density fiberboard	43.1		216.0		332.0	
Subtotal	300.6	32.7	564.0	42.5	851.0	54.1
Total	333.3		606.5		905.1	

fiberboard, core board, and plywood were 0.66, 1.08, 3.86, and 2.92 mg/m^2 h, respectively [19]. A survey in Beijing, Shanghai, and Guangdong indicated the passing rate of medium-density fiberboard, as far as the free formaldehyde emission rate was concerned, was only 61% [20].

Indoor pollution sources that release VOCs into the air are mainly solvent coatings for woodenware, interior architectural coatings, adhesives, wood-based furniture, carpets, and carpet cushions used in indoor decorating and refurbishing materials.

The emission rates of formaldehyde and VOCs from most indoor decorating and refurbishing materials are highest in new or renovated buildings. These compounds can "outgas" (i.e., evaporate continuously) slowly over months or years. As a consequence, the indoor concentration of these compounds slowly decreases, as shown in Tables 8 and 9. However, it can be also seen from

Table 8 Change in both indoor and outdoor formaldehyde concentrations after interior decoration in 15 Chinese households [21]

Time after indoor decoration	Less than 5 days	1 month	6 months	12 months
Indoor concentration range (mg/m^3)	0.852–13.402	0.095–1.002	0.082–0.561	0.009–0.129
Average indoor concentration (mg/m^3)	2.402	0.874	0.251	0.087
Average outdoor concentration (mg/m^3)	0.0065	0.0043	0.0044	0.007

Table 9 Changes in the concentrations of indoor aromatic compounds before and after interior decoration [9]

	Before decoration (mg/m^3)	During decoration (mg/m^3)	After decorating and refurbishing (mg/m^3)		
			10 days	31 days	12months
Benzene	0.024	0.164	0.109	0.054	0.037
Methybenzene	0.026	0.248	0.168	0.086	0.042
Ethylebenzene	Not detectable	0.105	0.075	0.031	0.019
1,2-Dimethylbenzene	0.012	0.153	0.121	0.051	0.026
1,3-Dimethylbenzene	Not detectable	0.090	0.069	0.025	0.018
1,4-Dimethylbenzene	0.009	0.320	0.187	0.079	0.036

Tables 8 and 9 that the indoor formaldehyde and VOCs concentrations are still much higher than the outdoor ones even in the 12th month after the indoor decoration project had finished [9, 21].

In recent years, there have been many complaints relevant to indoor air pollution caused by formaldehyde and VOCs emissions from decorating and refurbishing materials in China. Beijing City's Changping District People's Court judged China's first damages lawsuit for indoor air pollution in June 2001. Formaldehyde, in a concentration surpassing the indicated standard by 19.5 times, with value of 1.56 mg/m^3 in the owner's bedroom, resulted in the lawsuit [22].

2.3
Indoor Air Pollution from Building Materials

2.3.1
Ammonia

Indoor air quality can be reduced by ammonia released from building materials, furniture, cleaning compounds, office equipment, and other sources. At present, the acute health effects are associated with ammonia emitted from construction concrete in China. In mixing concrete, admixtures containing urea are added to improve resistance to damage from freeze – thaw cycles and to control such properties as setting time and plasticity. Later, the ammonia is released into the indoor environment from the concrete with an increase of environmental temperature, thus causing indoor air pollution. In 2000, severe indoor ammonia pollution occurred in Beijing. Some customers who bought apartments in a new housing development named Modern City felt strong ammonia irritation. Monitoring revealed that the ammonia concentration was about 20 times higher than the recommended limit (0.5 mg/m^3). It was confirmed that the ammonia came from construction concrete, in which a frost-resistant agent containing urea had been added during winter construction [23]. Ammonia was also implicated in the first foreign-related indoor air pollution lawsuit in March 2001, in Beijing. A domestic real-estate agency was prosecuted by a Beijing office of a foreign company for an indoor ammonia concentration as high as 8 mg/m^3 in a leased writing building [24].

2.3.2
Radon

Radon progeny – the decay products of radon gas – are a well-recognized cause of lung cancer in miners. When radon was found to be a ubiquitous indoor air pollutant, however, it raised a more widespread alarm for public health. Since 1994, a systematic radon survey has been done in 1,524 buildings and dwellings of 14 cities in China. The results showed that the highest indoor radon concen-

Table 10 Indoor radon level in some Chinese cities [25]

City	Sample number	Average value (Bq/m³)	Maximum value (Bq/m³)	>100		>200	
				Sample number	Percentage	Sample number	Percentage
Beijing	229	44.1	249	17	7.8	1	0.5
Qingdao	98	44.8	205	5	5.1	2	2.0
Taiyun	119	28.3	87.4	0	0.0	0	0.0
Bengbu	320	21.3	122	1	0.3	0	0.0
Lasha	44	44.6	125	1	2.3	0	0.0
Wuhan	56	25.7	170	1	1.8	0	0.0
Shangrao	150	82.1	596	38	25.3	6	4.0
Huanshan	12	49.5	96.6	0	0.0	0	0.0
Haikou	65	15.9	47.2	0	0.0	0	0.0
Guangzhuou	250	73.6	248	40	16.0	6	2.4
Shenzhen	189	35.3	332	2	1.1	1	0.5
Zhuhai	221	63.4	771	27	12.2	3	1.3
Zhengzhou	25	33.5	133	1	4.0	0	0.0
Pingliang	31	61.5	149	2	6.5	0	0.0
Total	1809			135	7.5	19	1.1

Table 11 Content of radioactive species of common building materials in China [27]

Building material	Natural stone	Brick	Cement	Sandrock	Lime	Soil
^{226}Ra	91	50	55	39	25	38
^{232}Th	95	50	35	47	7	55
^{40}K	1,037	700	176	573	35	584

tration is 596 Bq/m³ (Table 10) [25]. According to incomplete statistics, 50,000 people get lung cancer from radon in China every year [26].

The building materials emitting radon include stone, brick, soil, and sand. The radioactive species contents of common building materials in China are given in Table 11 [27]. The relatively high radioactive species are found in natural stone. Details of the radioactive species found in natural stone used in China are listed in Table 12 [28].

Besides the indoor air pollution caused by residential energy consumption, decorating and refurbishing materials, and building materials, indoor air pollutants also come from other sources, such as environmental tobacco smoking, products for household cleaning and maintenance, human metabolism, and outdoor contaminated air, as in any other country. See other chapters of this volume for discussions of the characteristics of indoor air pollution caused by these sources.

Table 12 Content of radioactive species of common natural stone materials in China [28]

Type	Rock character	^{226}Ra		^{232}Th		^{40}K	
		Average value	Range	Average value	Range	Average value	Range
Marble	Ultrabasic rock	25.2	0.37–97	11.9	0.59–193	105	9–1,003
	Basic rock	9.6	4.0–25.4	13.6	0.5–53.5	353	17–787
Granite	Neutral rock	52.1	20.9–155	69.6	3.2–201	941	281–1,618
	Acidic rock	79.6	6.3–374	99.9	9.8–276	1,128	446–1,810
	Alkali rock	126.9	53.7–200	158	65.8–252	2920	2,419–3,357
Meta-morphite	Meta-morphite	48.2	16.7–172	48.6	18.4–81.2	1,064	754–1,369
Slate	Slate	10.6		4.2		241	

3
Main Reasons for the Indoor Air Quality Problem

At present, indoor air pollution is already severe in China. This situation has resulted from many factors. The major reasons are the following:

1. The importance of indoor air quality is not recognized. On average, up to 80% of a person's day is spent at home and at the office, meaning the indoor air quality is of paramount importance to people's health. However, the public lacks knowledge about indoor air pollutants, their sources, characteristics, and health effects, because of insufficient environmental education. For this reason, the importance of maintaining a clean indoor environment is not widely appreciated.
2. The indoor environment administrations are not established. In China, many departments including environmental protection, occupational safety and health, and construction all deal with the indoor environmental problem to varying degrees. However, none of them are granted the authority or given the responsibility of managing indoor environmental quality. As a consequence, neither a special indoor environment supervisory agency nor a monitoring network has been established.
3. Indoor air pollution is not comprehensively investigated. Although a study on the indoor air pollution problem began in the early 1980s in China, only researchers from the fields of preventive medicine have been concerned with such a problem. Multidisciplinary research into the control of indoor air pollution has been conducted. On the other hand, there is no any agency that

has provided professional indoor air quality monitoring services for a long time in China. There are no systematic data about indoor air pollution and its heath effects that could be used to caution the public. A scientific system of evaluating indoor air quality has not been set up, and economically feasible technologies for controlling indoor air pollution are not available.

4. Disordered decorating and refurbishing markets. In China, high-speed economic development has brought prosperity to the decorating and refurbishing industry. Indoor decorating and refurbishing of buildings have become the consumption hotspot of urban and rural residents, which drives rapid growth of production and the use of indoor decorating and refurbishing materials. At the same time, some inferior products containing harmful substances also enter the market because of nonnormalized order.

5. Poor sanitation conditions in kitchens and bathrooms. Although the housing situation of urban and rural residents in China has greatly improved in recent years, per capita usable floor area still is low, especially in kitchens and bathrooms. In addition, ventilation throughout many households is poor. Serious air pollution and high moisture exist in kitchens and bathrooms.

6. Poor sanitation in public places. The population density in public places such as shopping centers, waiting rooms of hospitals and public transportation, and entertainment centers tends to be very high in China. Serious air pollution in these places occurs owing to high carbon dioxide, ammonia, hydrogen sulfide and pathogenic organism concentrations.

7. Air-conditioning systems are improperly maintained. Air conditioning is becoming more and more prevalent with the improvement of people's living conditions. At present, air-conditioning systems are installed in almost all modern office buildings and about 30% of urban households in China [29]. When air conditioning is in use, the building tends to be more tightly sealed, thus reducing natural indoor–outdoor air exchange. On the other hand, the air-conditioning systems can be the source of indoor air contamination or can act as the pathway through which other contaminants enter the airstream and are circulated throughout the building. Indeed, many air-conditioning systems are responsible for serious indoor air pollution due to dirt and moisture buildup caused by improper maintenance or the age of equipment.

4
Strategies for Reduction of Indoor Air Pollution

In order to control indoor air pollution, there have been a series of legal enactments, policies and measures which have been carried out since the 1980s in China, these have included the following.

4.1
Constituting Standards and Guidelines

As early as 1988, China published the first set of hygienic standards for public places, in which a limit on the amount of pollution allowed was given for carbon monoxide, inhalabe particulates, carbon dioxide, and bacteria. The standards played an important role in improving sanitation in public places and controlling the propagation of diseases. An amendment of the standards, issued by the General Administration of Quality Supervision, Inspection and Quarantine of the People's Republic of China, took effect in 1996, especially increasing the limit for formaldehyde emission in public places (Table 13).

In 1996 GB/T 16146-1995 "Standards for controlling radon concentration in dwellings" and GB/T 16127-1995 "Hygienic standard for formaldehyde in indoor air of house" were brought into effect after many years of investigation into indoor radon pollution and chemical pollution.

Coal consumption is one of the major causes of both indoor and outdoor air pollution. Although indoor coal combustion is the predominant source of indoor pollutants, outdoor pollutant concentrations also affect indoor levels, depending on the difference between indoor and outdoor concentrations and house ventilation. In the mid 1990s, air pollution in China was very serious owing to China's large consumption of coal as a fuel. Sulfur dioxide emissions in China rank first in the world, and carbon dioxide emissions rank second in the world, second to the USA. High ambient concentrations of sulfur dioxide, nitrogen oxides, and particulate matter occurred in China. Because of this situation, hygiene standards for carbon dioxide, inhalable particulate matter, nitrogen oxides, and sulfur dioxide in indoor air were issued by the General

Table 13 Hygiene standards for public places in China

Standard number	Standard name
GB 9663-1996	Hygiene standard for hotels
GB 9664-1996	Hygiene standard for public places of entertainment
GB 9665-1996	Hygiene standard for public bathrooms
GB 9666-1996	Hygiene standard for barber's shops and beauty shops
GB 9667-1996	Hygiene standard for swimming pools
GB 9668-1996	Hygiene standard for gymnasiums
GB 9669-1996	Hygiene standard for libraries, museums, art galleries and exhibitions
GB 9670-1996	Hygiene standard for shopping centers and bookstores
GB 9671-1996	Hygiene standard for hospital waiting rooms
GB 9672-1996	Hygiene standard for waiting rooms of public transit means of transportation
GB 9673-1996	Hygiene standard for public means of transportation
GB 16153-1996	Hygiene standard for dining rooms

Table 14 Hygiene standards for carbon dioxide, inhalable particulate matter, nitrogen oxides, and sulfur dioxide in indoor air

Standard number	Standard name
GB/T 17094-1997	Hygiene standard for carbon dioxide in indoor air
GB/T 17095-1997	Hygiene standard for inhalable particulate matter in indoor air
GB/T 17096-1997	Hygiene standard for nitrogen oxides in indoor air
GB/T 17097-1997	Hygiene standard for sulfur dioxide in indoor air

Table 15 Limits of harmful substances of indoor decorating and refurbishing materials

Standard number	Standard name
GB 18580-2001	Limit of formaldehyde emission of wood-based panels and finishing products
GB 18581-2001	Limit of harmful substances of solvent coatings for woodenware
GB 18582-2001	Limit of harmful substances of interior architectural coatings
GB 18583-2001	Limit of harmful substances of adhesives
GB 18584-2001	Limit of harmful substances of wood-based furniture
GB 18585-2001	Limit of harmful substances of wallpapers
GB 18586-2001	Limit of harmful substances of poly(vinyl chloride) floor coverings
GB 18587-2001	Limit of harmful substances emitted from carpets, carpet cushions, and adhesives
GB 18588-2001	Limit of ammonia emitted from concrete admixtures
GB 6566-2001	Limit of radionuclides in building materials

Administration of Quality Supervision, Inspection and Quarantine as well as by the Ministry of Health of the People's Republic of China in 1997 (Table 14).

The improvement of people's living standards in recent years has led to a surge in home decorating in China. But substandard decorating and refurbishing materials and lack of national standards in stemming the harmful substances have exacerbated indoor air pollution. In order to change this situation, a new set of national standards that cap the limit of harmful substances in interior decorating materials has been issued. The ten State Standards, issued by the State General Administration for Quality Supervision and Inspection and Quarantine, took effect on January 1, 2002 (Table 15).

The rules address homeowners' mounting complaints about pollution by explicitly limiting the content and intensity of harmful chemicals used in home improvement. They include formaldehyde, VOCs, and ammonia contained in interior architectural coatings, wood furniture, adhesives, and carpets.

The decorating and refurbishing material manufacturers, sales agents, builders, and decorators should fully comply with the standards to minimize the harm caused by indoor pollution. Production businesses are required to

Table 16 Limit of environmental pollutants of civil building engineering work. Type 1 buildings include households, hospitals, homes for the elderly, kindergartens, and classrooms. Type 2 buildings include office buildings, shopping centers, hotels, public places of entertainment, bookstores, libraries, barber's shops and beauty shops, gymnasiums, exhibitions, restaurants, and waiting rooms of public means of transportation

Pollutant	Type 1 building	Type 2 building
Radon (Bq/m^3)	≤200	≤400
Free formaldehyde (mg/m^3)	≤0.08	≤0.12
Benzene (mg/m^3)	≤0.09	≤0.09
Ammonia (mg/m^3)	≤0.2	≤0.5
Total volatile organic compounds (mg/m^3)	≤0.5	≤0.6

observe the standards immediately. The sale of any indoor decoration material that fails to meet State standards was outlawed from July 1, 2002.

As a result of the publication of the limits of the harmful substances in the form of national standards, consumers will have an authoritative reference when they address any disputes that arise from indoor decoration. Some specifications in the national standards, like those for VOCs in interior architectural coatings, conform to those standards in the European Union and the USA.

In addition, GB50325-2001 "Code for indoor environmental pollution control of civil building engineering", jointly issued by the State General Administration for Quality Supervision and Inspection and Quarantine as well as the Ministry of Construction of the People's Republic of China, took effect on January 1, 2002. The code stipulates indoor environmental pollutants have to be monitored when civil building engineering work is examined and accepted. Only buildings whose indoor environmental quality attains the demands given in Table 16 are acceptable.

At present, the indoor environmental quality standard is under discussion. On the whole, standards and regulations controlling indoor air pollution have basically formed through many years of endeavors in China.

4.2
Strengthening Management of Energy Consumption

4.2.1
Restructuring Energy Patterns and Developing Central Heating

Heating and cooking by means of small coal or biomass stoves are one of the major reasons leading to indoor air pollution. With regard to different kinds of heating systems, separate ones (with small coal stoves) in individual residence units have higher indoor air pollution than central heating systems in winter. So, optimizing energy structure and developing central heating are reliable

Table 17 Per capita residential energy consumption [2]

Year	Per capita annual average residential energy consumption						
	Total energy (kgce)	Coal (kg)	Electricity (kWh)	Kerosene (kg)	Liquefied petroleum gas (kg)	Natural gas (m³)	Gas (m³)
1980	97.7	118.0	10.7	1.0	0.4	0.2	1.4
1985	126.7	148.7	21.2	1.2	0.9	0.4	1.2
1990	139.2	147.1	42.4	0.9	1.4	1.6	2.5
1991	138.1	142.0	46.9	0.8	1.7	1.6	3.1
1992	133.4	126.1	54.6	0.7	2.0	1.8	4.4
1993	130.6	120.5	61.2	0.6	2.5	1.4	4.5
1994	129.3	109.5	72.7	0.6	3.2	1.7	6.3
1995	130.8	112.3	83.5	0.5	4.4	1.6	4.7
1996	145.5	118.3	93.1	0.5	5.8	1.6	3.9
1997	133.1	99.5	101.6	0.5	6.0	1.7	4.9
1998	115.9	71.5	106.6	0.5	6.2	1.9	6.0
1999	115.3	66.6	159.4	0.6	7.0	2.0	6.5

Table 18 Basic statistics on central heating supply in China [2]

Year	Volume supplied		Heating capacity	
	Steam (t)	Hot water (10^7 kJ)	Steam (t/h)	Hot water (10^6 W/h)
1997	20,604	62,661	65,207	69,539
1998	17,463	64,684	66,427	71,720
1999	22,169	69,771	70,146	80,591
2000	23,828	83,321	74,148	97,417

policies and measures for controlling both outdoor air pollution and indoor air pollution. In recent years liquefied petroleum gas, natural gas, and city coal gas have been widely used and central heating capacity is rapidly rising in China, thus greatly decreasing coal consumption (Tables 17, 18). Using liquefied petroleum gas, natural gas, and city coal gas as fuels will reduce sulfur dioxide, carbon dioxide, and dust emissions, thus mitigating indoor air pollution.

Here, it should be pointed out that the west–east gas pipeline project, one of the biggest construction projects in China's history, started in 2001 and will finish in 2004. The 4,000-km pipeline project will transport natural gas from Tarim Basin in Xinjiang, Qaidam Basin in Qinghai, and Erdos Basin to the Yangtze River delta region. It will supply gas mainly to the Yangtze delta region and other provinces along the pipeline, such as the provinces of Gansu, Shannxi, Henan, and Anhui. The west–east gas pipeline will turn the energy advantages of western regions into potential economic gains. It also plays an important role in preventing and controlling both indoor and outdoor air pollution.

4.2.2
High-Grade Conversion and Utilization of Biomass Energy

A great deal of biomass was used as fuel for cooking and heat supply in China. The traditional method of utilization of biomass energy has not been able to suit the demand, along with the rapid development of the rural economy, continuous improvement of the peasants' living quality, and the increasing attention to the environment.

The grand plan of firewood-saving stoves implemented from early 1980s, not only raised the utilization rate of biomass energy, diminished the shortage of energy in the countryside, but also improved indoor air quality. In addition, China has carried out extensive research and development on high-grade conversion plants and the application of biomass technology. Significant progress has been achieved in the aspects of gasification, liquefaction, and compact forming of biomass. The biomass gasifying plant has already been put onto the market and is used for cooking, wood drying, and heat supply. The technique of fluidized-bed pyrolytic gas for centralized supply is being tested. At present, there are many research units, factories, and companies engaged in development, demonstration, mass production, and service of biomass energy.

High-grade conversion and utilization of biomass energy is one of the Priority Projects for Development of the New and Renewable Energy in China. Its objectives are to speed up the use of the utilization technology of biomass energy, to develop highly efficient and direct-burning technology and the technology of compact solid forming, gasification, and liquefaction, to establish highly efficient industrial production technology and plants. The high-grade utilization volume of biomass energy is planned to reach 17×10^6 tons of standard coal by 2010.

4.2.3
Alleviating Indoor Air Pollution by All Types of Intervention

In order to alleviate indoor air pollution from heating and cooking, many interventions have been introduced in China. Some of them have been tested to be effective in reducing indoor air pollution (Table 19).

4.3
Strengthening Management of Civil Building Engineering Work

Statistics indicate that the average annual floor spaces of newly built residences in urban areas and rural areas were 4.5×10^8 and 6.46×10^8 m^2, respectively, during the Ninth 5-year Plan in China. It is forecasted that the total floor space of

Table 19 Interventions to alleviate indoor air pollution in China

Intervention type	Intervention examples
Technologies which aim at improved cooking/heating devices, improved fuels, or reduced need for heating	Better stove design Better ventilation Chemical treatment of some fuels, for example, coal Reduce the size of fuel pieces, for example, briquettes and pellets instead of large coal lumps Better insulation
Technologies aimed at improving the living environment	Partitions, walls, or screens in homes to separate cooking and sleeping/living areas Better ventilation or ducts and hoods to carry smoke and particulates outside the house
Behavioral change to reduce exposure and/or reduce smoke generation	Reduce the time spent in the kitchen/cooking area Keep lids on pots while cooking Proper stove maintenance and cleaning Push fuel (especially plant stalks) deeper into the stove so that less smoke "escapes" into the room Keep children away from the smoke

new residential buildings will be 5.7×10^9 m^2 between 2001 and 2005 [30]. However, per capita, residential area is very low, being only 10.3 and 24.8 m^2 in urban areas and rural areas, respectively, in 2000 because of the vast population [2]. It is without question that house buying, indoor decorating, and refurbishing of existing and new buildings will still be in great demand. So, China lays special emphasis on indoor environmental quality management of civil building engineering work. It is required that the construction administration is in charge of the supervision and administration of building and refurbishment quality. At the same time, a series of concrete measures are taken.

4.3.1
Strictly Implementing Code for Indoor Environmental Pollution Control of Civil Building Engineering Work in the Survey, Design, and Construction of Buildings

On the one hand, indoor air pollution control has to be comprehensively considered during the engineering survey, indoor ventilation design, decorating, and refurbishing design. On the other hand, construction inspectors should prevent buildings and refurbishing materials whose content and intensity of harmful chemicals surpass the limits stipulated in standards from entering construction fields in the construction phase. At the same time, the enforcement of the ISO 1400 environmental management system and cleaner production should be introduced and encouraged during the survey, design, and construction.

4.3.2
Establishing an Inspection and Acceptance System for Indoor Environmental Quality Before Commissioning Civil Building Engineering Work

According to the demands of the code for indoor environmental pollution of civil building engineering work, an engineering construction unit is obligated to commission an authorized agency to monitor the contents of radon, formaldehyde, benzene, ammonia, and total volatile organic compounds in indoor air before commissioning civil building engineering work. Only when the indoor environmental quality meets the requirements of the code can the building engineering work begin.

4.3.3
Strengthening Supervision and Management for Indoor Environmental Quality of Civil Building Engineering Work

Supervision agencies for building engineering work have to regard indoor environmental quality as one of the major supervision contents for building engineering work. In engineering supervision reports submitted to administrations, the last comments on the indoor environmental quality of building engineering work have to be included. The official records are not done if the indoor environmental quality of the building fails to meet the requirements of the code.

4.3.4
Preventing Fake and Inferior-Quality Building and Refurbishing Materials from Entering the Market

In the past few years, fake and inferior-quality building, decorating, and refurbishing materials have been produced and sold by those who are blinded by gain because of nonnormalized market order, thus causing severe indoor air pollution. Because of this situation, spot checks for the building and refurbishing material market are routinely carried out. The production, sale, and use of any building and refurbishing materials that fail to meet state standards will be strictly forbidden and punished in order to ensure the quality of building, decorating, and refurbishing.

4.3.5
Neatening Quality Certification Order

At present, there are many quality certification agencies that provide certification services on building and refurbishing materials in China. However, the certificate market is in confusion because it lacks unified management and coordination between different administrations. For this reason, the Certification and Accreditation Administration of the People's Republic of China, together with other relevant administrations, will neaten the certification market and struggle against illegal behavior, thus creating a positive and ordered quality certification environment.

4.4
Other Strategies

4.4.1
Establishing a Lead Enforcement Agency for Managing Indoor Environments

In order to change the disordered state of the management of indoor environmental quality, the National Environmental Protection Agency has been designated as the lead agency for indoor environmental management. It was required to coordinate with other relevant agencies, such as the Department of Construction and the Ministry of Health, to address indoor air pollution problem.

4.4.2
Providing a Service of Indoor Air Monitoring

Citizens need an agency to turn to when they feel that their home or office has a contamination problem. To respond to these concerns, many agencies which provide indoor environmental monitoring services were organized in research institutions and universities in the past 3 years. A national center for indoor environmental monitoring was established. Training on monitoring of indoor pollutants was held many times. At the same time, special mobile monitoring instruments have been equipped to meet the requirement of in situ monitoring.

At present, examination and certification of indoor environmental monitoring agencies are under discussion. It has been affirmed that a normalized indoor environmental monitoring market, in a not-to-distant future, will be established in China, and it will play an important role in ensuring the accuracy of monitoring data.

4.4.3
Conducting Research on Indoor Air Pollution and its Health Effects

In order to gain an insight into the current situation of indoor air pollution, many research programs are being conducted in China. Research contents include (1) identifying the magnitude as well as the major sources of indoor air pollution, (2) identifying key health problems, (3) pollution control consistent with energy conservation, and (4) assessment of the risk of exposure to indoor air pollutants. Especially, there are three projects on indoor environment pollution and its control that are listed as the Key Technologies R&D Program in the Tenth 5-year Plan of the Ministry of Science and Technology of the People's Republic of China. They are health evaluation technologies of key indoor air pollutants, control technologies of key indoor air pollutants, and control technologies of indoor air pollution from coal burning. In addition, academic conferences on indoor air pollution and its health effects have been held many times in recent years. Multidisciplinary collaboration in research is strengthening.

4.4.4
Developing Effective Pollution Elimination and Control Technologies

At present, there are two different research interests in developing indoor air pollution elimination and control technologies in China:

1. Environmental benign building, decorating, and refurbishing materials. Manufacturing processes for building, decorating, and refurbishing materials are relatively out of date in China. To improve production processes, especially the development and use of environmental benign building, decorating, and refurbishing materials, is potentially simpler and a more effective contaminant mitigation measure than those measures that focus on removing contaminants after they become airborne or become entrained in indoor air. As a source control strategy, it will prevent or exclude the entry of formaldehyde, VOCs, radon, and ammonia, into building spaces.
2. Indoor air pollution control appliances. Indoor air pollution control is currently one of major interests of research and development in China. Many technologies for purifying indoor air, including adsorption, photocatalysis, catalysis, and plasma, are being investigated. Purification mechanisms, applicability, and factors affecting purification efficiency have been established. On the basis of these studies, indoor air purifiers and air-conditioning systems with high efficiency indoor air handling units have been developed and have been accepted by consumers.

4.4.5
Increasing Public Awareness of the Importance of Indoor Air Quality

A key factor in reducing indoor air pollution is an improvement in public information. To increase public awareness of how individual activities and consumer choices affect the environment could make cleanup efforts more successful. More public information is also helpful to keep the public interested in current environmental issues and to foster their sense of responsibility to work for a better and greener world. It would also increase public pressure for pollution reforms, which would in turn leverage more money from the government for emission controls and environmental cleanups.

In recent years dissemination of indoor environmental knowledge and expertise has been greatly promoted in China. The general public has been educated about the causes and effects of pollution, especially health hazards from certain decorating and refurbishing materials, combustion products, furnishings, construction and maintenance, pesticides, home and office products, and appliances. At the same time, instruction in the correct use and maintenance of products, substitution of nonpolluting products, and recommended levels and frequency of ventilation are provided to the public. After having had insight into indoor air pollution and its health effects, the public is not only making more rational decisions in decorating households, selecting appliances, using air-conditioning systems, etc., but is also urging the government to take a series of measures to control indoor air pollution.

5
Concluding Recommendations

The emission of various air pollutants into indoor environments brings about severe indoor air pollution. There is a strong desire to regulate and reduce the levels of these pollutants. To address the challenges, we note some policy problems and make the following concluding recommendations.

First, poor quality fuels and unventilated stoves will continue to be largely used in rural areas though enormous advances in the optimization of energy patterns and development of central heating systems in urban areas have achieved in the past few years. So, structural reform and rational use of fuels, improvement of stoves and, ventilation systems will play key roles in controlling indoor air pollution in rural areas.

Second, it is clear that house buying and indoor decoration and refurbishment of existing and new buildings will still be in great demand in view of the economic development trend and the current living space situation in China, especially in urban areas. Managing the construction and indoor decoration and refurbishment of the residential buildings will continue to be one of the most critical factors in the control of indoor air pollution in China.

Last, we successfully draw lessons from international experience in controlling outdoor air pollution. In the same way, it is very necessary to strengthen cooperation with developed countries, and to make use of their experience and technology to solve indoor air pollution problems because these countries suffered from similar problem during their periods of rapid development.

References

1. http://www.dtdjc.com/jjzs/2002–4-7/wgjz.htm
2. State Statistical Bureau (2001) China statistical yearbook. China Statistical Press, Beijing
3. Shen JM (1996) PhD thesis, Tongji University
4. Liu YJ, Zhu LZ, Shen XY (2001) Environ Sci Technol 35:840
5. Florig HK (1997) Environ Sci Technol 31:276A
6. Zhang LF, Smith KR (1999) Environ Sci Technol 33:2311
7. Zhang LJ Goldberg MS Gao YT Jin F (1999) Cancer Causes Control 10:607
8. Zhang L (1998) PhD thesis. Nankai University

9. Jing YL (2002) Indoor air pollution and health. In: Proceedings of the 1st national symposium on indoor air quality and health. The Chinese Preventive Medicine Society, Beijing, p 16

10. Qu JQ, Wu SA (2002) Indoor environmental monitoring technology. In: Training materials on indoor environment and health. The Chinese Society of Environmental Science, The Chinese Expert Committee on Indoor Environment and Health, Beijing, p 1

11. http://lnweb18.worldbank.org/essd/essd.nsf/EnvironmentStrategy/Brochure-home

12. Pang XC, Dong ZZ, Jin XB, Wang BY, Wang LH, Xu XP (2002) Indoor and outdoor air pollution and their effects on resporatory system in western Anhui' rural area. In: Proceedings of the 1st national symposium on indoor air quality and health. The Chinese Preventive Medicine Society, Beijing, p 190

13. Peng RC, He KB, Wang LH, Xu XP, Wang H (1998) In: McElroy MB, Nielsen CP, Lydon P (eds) Energizing China: reconciling environmental protection and economic growth. Harvard University Committee on Environment, MA, p 287

14. http://www.igc.org/wri/wr-98-99/prc-air.htm

15. Zhong LJ, Goldberg MS, Gao YT, Jin F (1999) Epidemiology 10:488

16. Gao HB, Zhang DG, Liu FC (1999) Occup Med 26:17

17. http://www.chinaenvironment.net/sino/sino4/page22.html

18. http://www.srs.fs.fed.us/pubs/rpc/1999-09/rpc_99sep_20.pdf

19. Song RJ (2002) Emission rate of formadelhyde from wood-based panels. In: Training materials on indoor environment and health. The Chinese Society of Environmental Science, The Chinese Expert Committee on Indoor Environment and Health, Beijing, p 85

20. http://www.gdsnhj.com/news/f/f003.html

21. Dai TY, Wei FS, Liu DQ (2001) Indoor air pollution in decorated houses. In: Proceedings of 2001 international symposium on indoor air quality. National Environmental Protection Agency of China, Beijing, p 1

22. http://www.snhj.net/update/sjds.htm

23. http://house.enorth.com.cn/zycm/alfx/000146862.html

24. http://www.legaldaily.com.cn/gb/content/2001-04/11/content_16167.htm

25. Qi QP, Xu DP, Zhu XS, Dai ZZ, Chen XP (2002) A study on indoor air quality standard. In: Proceedings of the 1st national symposium on indoor air quality and health. The Chinese Preventive Medicine Society, Beijing, p 6

26. http://rich.online.sh.cn/rich/gb/content/2001–08/24/content_202898.htm

27. Wang ZY (2002) Residential radioactivity on health. In: Proceedings of the 1st national symposium on indoor air quality and health. The Chinese Preventive Medicine Society, Beijing, p 12

28. Hang YZ (2002) Radioactivity of building materials and radon monitoring. In: Training materials on indoor environment and health. The Chinese Society of Environmental Science, The Chinese Expert Committee on Indoor Environment and Health, Beijing, p 50

29. http://finance.sina.com.cn/x/20011205/151224.html

30. http://www.cbh-jj.com/page/sexx.htm

Subject Index

Printing: Strauss GmbH, Mörlenbach
Binding: Schäffer, Grünstadt